IoT and OT Security Handbook

Assess risks, manage vulnerabilities, and monitor threats
with Microsoft Defender for IoT

Smita Jain

Vasantha Lakshmi

BIRMINGHAM—MUMBAI

IoT and OT Security Handbook

Group Product Manager: Vijin Boricha

Publishing Product Manager: Mohd Riyan Khan

Senior Editor: Runcil Rebello

Technical Editor: Arjun Varma

Copy Editor: Safis Editing

Language Support Editor: Safis Editing

Project Coordinator: Ashwin Kharwa

Proofreader: Safis Editing

Indexer: Sejal Dsilva

Production Designer: Nilesh Mohite

Marketing Coordinator: Marylou De Mello

First published: March 2023

Production reference: 1100323

Published by Packt Publishing Ltd.

Livery Place

35 Livery Street

Birmingham

B3 2PB, UK.

ISBN 978-1-80461-980-3

www.packtpub.com

I have to start by thanking my mother, Sneh lata, my mother-in-law, Madhu Shah, and my late father, Shree Manohar lal, for their blessings. Also, my darling daughter, Neel Jain, and my husband, Pratik Jain – for being a supporter of my idea, giving me space and time to follow my passion for cybersecurity and when writing the book, and keeping the munchkin out of my hair so I could edit. Thank you so much to my coauthor, Vasantha, for giving me the opportunity to fulfill a dream and bring this book to completion. Without her help, this book would have been just a dream.

– Smita Jain

To my mother, Geetha Sabhahit, my husband, Muralidhar K., and my lovely baby girl, Dyuthi Chandahsara, for being my support and the rhythm of my life. To everybody I interacted with and have learned something from throughout my life! And to Smita Jain, may our partnership continue to prosper.

– Vasantha Lakshmi

Foreword

The age of Industry 4.0 has brought about a rapid increase in the use of IoT and OT devices across various industries. While these devices have transformed the way we work and interact with each other, they have also brought about new cybersecurity challenges that must be addressed.

This book on IoT and OT security provides a timely and comprehensive resource for anyone looking to enhance their understanding of these challenges. The book is divided into three parts, each of which provides unique insights into different aspects of IoT and OT security.

The first part offers readers an in-depth understanding of the security challenges in IoT and OT environments, including common attacks and vulnerabilities. This section lays the foundation for the rest of the book, offering you a comprehensive understanding of the risks and threats associated with the use of IoT and OT devices in Industry 4.0.

The second part focuses on Microsoft Defender for IoT and how it can help address the open challenges in the connected world we live in today. The authors provide practical guidance on deploying and using Microsoft Defender for IoT to secure IoT and OT environments. This section offers you a comprehensive approach to securing your devices against cyber threats.

The third part offers best practices for achieving continuous monitoring, vulnerability management, threat monitoring, and hunting in IoT and OT environments. It highlights the importance of risk assessment and aligning the business model towards Zero Trust, providing you with a holistic approach to securing your IoT and OT devices against cyberattacks.

In conclusion, this book is a valuable resource for anyone looking to enhance their knowledge of IoT and OT security in the age of Industry 4.0. It provides you with a comprehensive understanding of the challenges and risks associated with these environments and offers practical guidance on how to secure them. Whether you are an industry professional or cybersecurity expert, this book is an essential addition to your reading list.

Dr Rohini Srivathsa

National Technology Officer (CTO) - Microsoft India

Contributors

About the authors

Being a business enabler and risk management-focused, **Smita Jain** has a unique, globally experienced approach to information security, data privacy, IT, OT/IoT, and digital transformation.

Her hallmarks include transformative vision casting and strategy setting, operational and organizational excellence, and a risk-based approach to enterprise enablement.

Smita is recognized as a leader in the transformation process, re-envisioning and establishing organizational cadence and cultures, with an established track record of effectively working across various industries.

Smita is a cybersecurity consultant, strategist, and mentor to organizations in the ever-changing cybersecurity landscape, helping them build a dynamic cybersecurity program.

Vasantha Lakshmi works at Microsoft India as a technology specialist and was previously a program manager. She holds the **Certified Information System Security Professional (CISSP)** certification, which aids in aligning with the industry standard of security. She has worked on various security products for the last 7 years. She has more than 12 years of experience working as an architect of end-to-end cybersecurity solutions (for devices, data, apps, O365, identity, and so on) for Microsoft 365. Her Prosci Certified Change Practitioner certification has also aided her on her journey to digitally transform organizations. She holds many other certifications, such as M365 Enterprise Administrator Expert, M365 Desktop Administrator Associate, SC-200, SC-300, and MS-500.

About the reviewers

Abbas Kudrati, a long-time cybersecurity practitioner and CISO, is Microsoft Asia's lead chief cybersecurity advisor for the security solutions area. In addition to his work at Microsoft, he serves as an executive advisor to LaTrobe University, HITRUST Asia, EC-Council Asia, and several security and technology start-ups. He supports the broader security community through his work with ISACA chapters and student mentorship.

He is the bestselling author of books such as *Threat Hunting in the Cloud*, *Zero Trust Journey Across the Digital Estate*, and *Managing Risks in Digital Transformation*.

He is also a part-time Professor of Practice with LaTrobe University and a keynote speaker on zero-trust, cybersecurity, cloud security, governance, risk, and compliance.

Table of Contents

3

Common Attacks on IoT/OT Environments 21

Part 2: How Microsoft Defender for IoT Can Address the Open Challenges in the Connected World We Live in Today

4

What Is Microsoft Defender for IoT? 35

5

How Does Microsoft Defender for IoT Fit into Your OT/IoT Environment/Architecture? 47

6

How Do the Microsoft Defender for IoT Features Help in Addressing Open Challenges? 63

Part 3: Best Practices to Achieve Continuous Monitoring, Vulnerability Management, Threat Monitoring and Hunting, and to Align the Business Model Toward Zero Trust

7

Asset Inventory 93

Preface

This book gives a quick overview of **internet of things** (**IoT**) and **operational technology** (**OT**) architecture and then dives into securing IoT and OT devices. You will learn about **Microsoft Defender for IoT** (**MDIoT**) to secure IoT and OT devices against cyberattacks.

Who this book is for

This book is for CEOs, DEOs, CDOs, CFOs, B-CISOs, operations teams, enablers of the fourth Industrial Revolution (Industry 4.0), Zero-Trust/SASE enablers, heads of IT, digital transformation leaders, OT operations, security operations, and so on.

What this book covers

Chapter 1, *Addressing Cybersecurity in the Age of Industry 4.0*, covers Industry 4.0, which is the digital transformation that the manufacturing and production industries are going through in the connected world we live in today. However, there are a number of cybersecurity challenges associated with it, and they need immediate attention.

Chapter 2, *Delving into Network Segmentation-Based Reference Architecture – The Purdue Model*, explains the Purdue model and the industrial network architecture for **industrial internet of things** (**IIoT**), IoT, OT, **supervisory control and data acquisition** (**SCADA**), **integrated computer systems** (**ICS**), and more. OT is the base of the Purdue model.

Chapter 3, *Common Attacks on IoT/OT Environments*, elaborates on some of the common attack vectors that we see in IoT/OT. We will see how some of these attacks were made due to the weakness in the system.

Chapter 4, *What is Microsoft Defender for IoT?*, defines **MDIoT**, its features, and capabilities.

Chapter 5, *How Does Microsoft Defender for IoT Fit into Your OT/IoT Environment/Architecture?*, discusses multiple architectures and use cases and how we can fit MDIoT in various environments.

Chapter 6, *How Do the Microsoft Defender for IoT Features Help in Addressing Open Challenges?*, enumerates some of the business challenges of managing and securing IoT devices. Certain challenges are particularly important to address today as they affect not only business but also human life.

Chapter 7, *Asset Inventory*, builds upon the belief that your assets are of paramount importance, so we will show how increasing visibility into your assets will help reduce risk.

Chapter 8, Continuous Monitoring, helps with the detection of policy violations, industrial malware, anomalies, and operational incidents.

Chapter 9, Vulnerability Management and Threat Monitoring, looks at risk assessment, which forms an integral part of MDIoT. For example, the top vulnerable devices, variation from the baselines, remediation priorities based on the security score, network security risks, illegal traffic by firewall rules, connections to ICS networks, internet connections, access points, industrial malware, indicators, unauthorized assets, weak firewall rules, network operations, protocol problems, backup servers, disconnections, IP networks, protocol data volumes, and attack vectors are some of the topics discussed.

Chapter 10, Zero Trust Architecture and the NIST Cybersecurity Framework, focuses on Zero Trust, a security imperative. IoT devices, especially, are highly attacked, paving the way into the internal network and hence the OT network as well. MDIoT, with a minimal **operating system (OS)** footprint, can help organizations with rich information on risky **OS** configurations, strong identity for authentication, and so on to determine anomalies and unauthorized activities. Enterprise IoT is also discussed.

To get the most out of this book

It is best if you work in or know about Industry 4.0 and its cybersecurity challenges, or you are most welcome to learn about it from the book itself.

Software/hardware covered in the book	Operating system requirements
Defender for IoT physical appliance for sensor	NA
Defender for IoT virtual appliance on a **virtual machine (VM)**	The sensor provided is Linux and can be set up on Hyper-V or VMware VMs

Download the color images

We also provide a PDF file that has color images of the screenshots and diagrams used in this book. You can download it here: `https://packt.link/O5D0H`

Conventions used

There are a number of text conventions used throughout this book.

`Code in text`: Indicates code words in text, database table names, folder names, filenames, file extensions, pathnames, dummy URLs, user input, and Twitter handles. Here is an example: "Select `eth0` (or `eth1` based on your configuration)."

A block of code is set as follows:

```
#include <Windows.h>
Int main(void) {
MessageBoxA(0, "hi there.", "info", 0);
return 0;
}
```

When we wish to draw your attention to a particular part of a code block, the relevant lines or items are set in bold:

```
#include <Windows.h>
Int main(void) {
MessageBoxA(0, "hi there.", "info", 0);
return 0;
}
```

Any command-line input or output is written as follows:

```
$ mkdir css
$ cd css
```

Bold: Indicates a new term, an important word, or words that you see onscreen. For instance, words in menus or dialog boxes appear in **bold**. Here is an example: "Go to the **Plans and Pricing** section situated on the left under the **Management** header."

> **Tips or important notes**
> Appear like this.

Get in touch

Feedback from our readers is always welcome.

General feedback: If you have questions about any aspect of this book, email us at customercare@packtpub.com and mention the book title in the subject of your message.

Errata: Although we have taken every care to ensure the accuracy of our content, mistakes do happen. If you have found a mistake in this book, we would be grateful if you would report this to us. Please visit www.packtpub.com/support/errata and fill in the form.

Piracy: If you come across any illegal copies of our works in any form on the internet, we would be grateful if you would provide us with the location address or website name. Please contact us at `copyright@packt.com` with a link to the material.

If you are interested in becoming an author: If there is a topic that you have expertise in and you are interested in either writing or contributing to a book, please visit `authors.packtpub.com`.

Share Your Thoughts

Once you've read *IoT and OT Security Handbook*, we'd love to hear your thoughts! Scan the QR code below to go straight to the Amazon review page for this book and share your feedback.

`https://packt.link/r/1804619809`

Your review is important to us and the tech community and will help us make sure we're delivering excellent quality content.

Download a free PDF copy of this book

Thanks for purchasing this book!

Do you like to read on the go but are unable to carry your print books everywhere?

Is your eBook purchase not compatible with the device of your choice?

Don't worry, now with every Packt book you get a DRM-free PDF version of that book at no cost.

Read anywhere, any place, on any device. Search, copy, and paste code from your favorite technical books directly into your application.

The perks don't stop there, you can get exclusive access to discounts, newsletters, and great free content in your inbox daily

Follow these simple steps to get the benefits:

1. Scan the QR code or visit the link below

https://packt.link/free-ebook/9781804619803

2. Submit your proof of purchase
3. That's it! We'll send your free PDF and other benefits to your email directly

Part 1: Understand the Challenges in IoT/OT Security and Common Attacks

In this section, you will start by understanding the importance of Industry 4.0 and its impact on cybersecurity, along with some of the common attacks on the OT environment. You will then delve deeper to get an insight into network segmentation through the Purdue model and how it aids in securing an organization.

This section has the following chapters:

- *Chapter 1, Addressing Cybersecurity in the Age of Industry 4.0*
- *Chapter 2, Delving into Network Segmentation-Based Reference Architecture – the Purdue Model*
- *Chapter 3, Common Attacks on IoT/OT Environments*

1

Addressing Cybersecurity in the Age of Industry 4.0

We are seeing fast-paced digital transformation in all industries, including the **operational technology (OT)** and **Internet of Things (IoT)** industries.

The different eras of industry, as seen in *Figure 1.1*, have brought significant changes in the ways businesses work. Revolutionary changes within the way that industries operate have been brought about by manufacturing companies becoming more secure, efficient, productive, and profitable:

Industrial Revolution

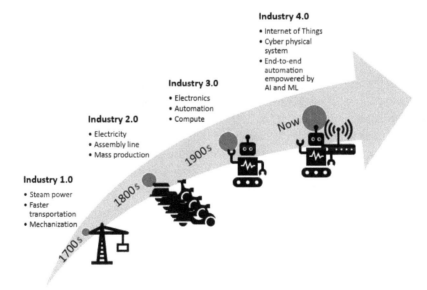

Figure 1.1 – The different eras of industrial revolution

The computing technology of Industry 4.0 is powered by a connected network and has a digital twin on the internet too. This enables communication and information sharing. This networking is sometimes referred to as a cyber-physical production system. This makes organizations and industries more autonomous. First, let us look at how Industry 4.0 came to be:

- **Industry 1.0**: innovations of steam power and mechanical engines in Industry 1.0 led to the faster conveyance of goods and people, resulting in enormous time savings.

- **Industry 2.0**: Electrical power and electronic assembly brought down the cost of production and business became more profitable and agile to demands.

- **Industry 3.0**: With computers and programming logic, businesses started automating complete assembly lines; so, parts were built faster as the assembly line system became more efficient at performing tasks. However, during this revolution, things remained human-driven and machine-executed. The fast pace amazed consumers and businesses.

- **Industry 4.0**: This revolution was built on top of Industry 3.0, in which isolated devices were connected through computer networks. Meanwhile, manually driven operations were converted into fully automated and integrated systems with digital twins on the internet to simulate different tasks on devices. The entire field was positively impacted by all the insights incorporated from demand to supply. Connected devices brought more intel.

The entire journey from Industry 1.0 to 4.0 has made business more profitable, agile, and cost-effective. Automation and intelligence have also introduced visibility into the demand and supply of raw materials and enhanced the quality of finished products. Businesses are now more empowered by all the intelligence derived from the system to make the right decisions.

Microsoft presently invests over $5 billion in IoT research and is a pacesetter regarding this new technology, thus contributing to the overall Industry 4.0 revolution. **Microsoft Defender for IoT (MDIoT)** is accelerating this digital transformation for organizations, with comprehensive security across IoT or OT infrastructure. MDIoT provides agentless **network detection and response (NDR)**. The technology is rapidly deployed and works with various IoT, OT, and **industrial control systems (ICS)**. For IoT device producers, MDIoT provides a lightweight agent to enhance device-level security. It is a solution that interacts with Microsoft 365 Defender, Microsoft Sentinel, Microsoft Defender for Endpoint, devices, and external **security operations center (SOC)** tools. It can be deployed on-premises, in a hybrid setup, or via the cloud. MDIOT covers the IoT, OT, and the **Enterprise Internet of Things (EIoT)**. This book focuses on securing OT.

Before we get to that, though, in this chapter, we will cover the following topics:

- How is Industry 4.0 being leveraged?

- Understanding cybersecurity challenges in the age of Industry 4.0

- Enumerating the factors influencing IoT/OT security

- How to overcome security challenges

How is Industry 4.0 being leveraged?

The power of **artificial intelligence** (**AI**) tools has led to the evolution of Industry 4.0. Various sectors—such as manufacturing, healthcare, finance, the public sector, consumer goods, retail, and smart city planning and building—use intelligent machines to help **subject matter experts** (**SMEs**) streamline processes, thus helping to meet highly competitive industry demands. How is this leveraged by different industry verticals?

The following figure describes how Industry 4.0 is leveraged by different industry verticals:

Figure 1.2 – Industries leveraging industry 4.0

The factors that are common across all verticals are the reduced cost of operations and enhanced customer experience, ultimately making companies more secure, efficient, productive, and profitable.

Now that we have seen some of the verticals leveraging and benefiting from Industry 4.0, we can go into detail regarding how these industries operate in the real world. In the real world, the convergence of **information technology** (**IT**) and OT brings about new challenges to conquer – that is, cybersecurity – and this convergence is bound to happen in pursuit of better business results, profits, data analytics, and much more.

Let us quickly learn a few basic terms:

- OT is hardware and software that detects or causes a change through directly monitoring and/or controlling industrial equipment, assets, processes, and events.

- The IoT is the concept of interconnected computer devices, mechanical and digital machines, objects, animals, or people, armed with **unique identifiers** (**UIDs**) and the ability to transmit data over a network, without intervention or computer interaction.

- The EIoT refers to how digital transformation and enterprise strategies being connected enables seamless collaboration between people and technology. It also provides more insight and improves the productivity of an enterprise by reducing the amount of manual work and optimizing business outcomes.

- The **Industrial Internet of Things (IIoT)** refers to the expansion and use of the IoT in industrial applications. It focuses strongly on **machine-to-machine (M2M)** communication, big data, and **machine learning (ML)**. The IIoT enables industries and businesses to make their operations more efficient and reliable. It encompasses industrial applications, including robotics, medical devices, and software-defined manufacturing processes.

- **Supervisory control and data acquisition** (SCADA) is a control system architecture that includes computers, networked data communications, and graphical user interfaces to monitor machinery and processes at a high level. This includes sensors and other devices, such as **programmable logic controllers (PLCs)**, that are connected to process systems or machines.

- A **distributed control system (DCS)** is an automated control system for a process or plant, usually with many control loops, in which self-contained controls are distributed throughout the system, but there is no centralized operator supervision. This is in contrast to systems with centralized controllers, discrete controllers in a central control room, or a central computer. The DCS concept increases reliability and reduces installation costs by placing control functions close to the process installation with remote monitoring.

The following figure provides more insight into the convergence of IT and OT, along with the IoT:

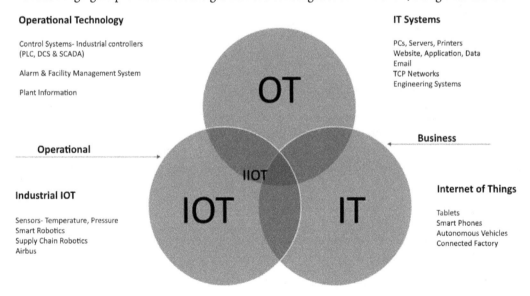

Figure 1.3 – Convergence of IT, OT, and the IoT

While OT covers control systems such as PLCs and DCS, IT covers PCs, servers, websites, and applications hosted within the organization or leveraged by an organization. The IT industry focuses on data while OT is focused on operations and the IoT focuses on M2M communication over the internet.

The IIoT industry, on the other hand, is about the use of smart sensors and actuators to enhance the outcome of industrial processes. This represents the intersection between OT and the IoT.

The legacy model of operation for organizations required IT and OT to be isolated. However, the convergence of IT and OT has brought about a successful transformation for the OT and IoT industries.

The IoT has already crossed the nascent stage and it has become mainstream. In fact, the number of IoT devices has already surpassed the number of IT devices. This proliferation of IoT devices has brought about increased attention, and adversaries and threat actors have already started targeting them for quick monetary gain by exploiting their vulnerabilities.

Industry 4.0 has three important stages:

1. **Digitization**: Digitized outputs that are connected to industrial assets lead to the convergence of the digital and physical worlds. Connecting digitized outputs to **security information and event management** (**SIEM**) solutions can provide visibility to all incidents and actions in real time.

2. **Sensorization**: Adding sensors to industry processes helps bring about better interconnection. This enables the autodetection of issues or changes in temperature, pressure, humidity, and so on, as it helps ascertain that the continuous monitoring of inputs and any sudden changes to them is in place. This prevents mishaps, too.

3. **Optimization**: With the AI and ML world we live in, data analytics is leveraged to improve business and process outcomes. Leveraging data analytics and simulations leads to optimized results, which bring about profits for the organization through the management of time and effort.

So far, we have understood the impact of and the blazing changes brought in by Industry 4.0, and how a connected world and its disruptive technology (the use of AI and ML) have changed our future trajectory in general.

Understanding cybersecurity challenges in the age of Industry 4.0

It is a known fact that cybersecurity breaches in any industry negatively affect business outcomes, and this is still a concern in the age of Industry 4.0. Cyberattacks on critical industrial equipment hamper businesses. Digitization is driven by four kinds of disruption:

- **The astronomical rise in data volumes**: As everything moves toward digitization, we can imagine the wealth of data generated as a result, thus making it easy for an attacker to intercept data and steal **intellectual property** (**IP**) as the amount of data that needs to be protected against threats and malware increases.

- **Forever-increasing computational power**: The job of attackers gets easier every time we see an increase in computational power. Quantum computing has increased our computational power manifold. This has empowered attackers to break or crack open more boundaries and types of authentication faster.

- **A super-connected world (connected vehicles, airbuses, heart defibrillators, etc.):** The attack surface has increased to an extent that it can now affect human life as well. Thinning down network segmentation, collecting data from sensors, and sending it for further processing have contributed to an increased attack surface.

- **A data (AI and ML) world (deep insights, business intelligence, and analytics):** Analytics are key for any organization to thrive today. While data brings about more business intelligence and provides opportunities for businesses to take calculated steps toward success, if infiltrated, these insights and intelligence fall directly into the hands of an attacker.

As a lot of industries adapt to Industry 4.0, they have become an appealing target for attackers. Attackers jump across IT and OT laterally on the lookout for vulnerabilities to exploit them, paving the way for industrial espionage, IP leakage, or IP theft.

The improved connectivity between and the convergence of OT and IT have increased the attack surface. This also requires that the compliance requirements of every industry be retouched to meet the ever-growing security requirements.

We are here to address this challenge and aid organizations in line with Industry 4.0 to protect their infrastructure from attacks continuously. We can do this with **MDIoT**. We will see more about MDIoT in upcoming chapters.

Cybersecurity challenges are ever-increasing, and the few that affect the IoT/OT world that we have mentioned are just examples to get us started on this journey to understand the challenges and how to address them better with MDIoT.

Enumerating the factors influencing IoT/OT security

Attacks on internet devices have increased manifold in recent years and are commonplace now. The reasons for this are the following:

- **Weak credentials:** How often have we ignored notifications to change our default passwords? We cannot forget the *Mirai botnet* attack, which made use of default usernames and passwords of this kind.

- **Legacy software and hardware:** We don't always see IoT devices being updated. With the pace at which technology is growing, it becomes very easy for attackers to find weaknesses. Hence, it is of paramount importance to always update devices.

- **Proprietary protocols and the complex structure of OT devices:** With so many types of devices in use, we can only imagine the many proprietary protocols and the complex structures they would bring in. Siemens, Schneider, ABB, Rockwell Automation, and so on all use closed protocols.

- **Anomalous operational events:** Operations such as start, stop, and restart are all priority commands for **Retentive Timers (RTOs)**, PLCs, actuators, and so on. Hence, it becomes a priority to clearly distinguish legitimate events from anomalous events.

These are some of the obvious factors that affect IoT/OT security. As we proceed further through the book, we will see more of these factors and how MDIoT addresses the challenges involved.

How to overcome security challenges

IoT/OT security should focus on providing visibility into the landscape of connected devices within an organization while going by the principle, *"If you do not need the device at all, do not have it!"* Reduce your attack surface.

However, if you do need a device, ensure that you have an accurate asset inventory. If you do not know the assets owned by your organization, you do not know what to protect—especially in a decentralized environment with multiple IoT/OT devices:

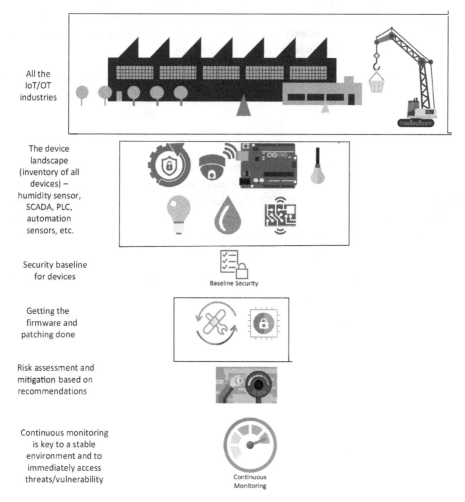

Figure 1.4 – Approach to securing an IoT/OT infrastructure

A **trusted computing base** (**TCB**) or a security baseline approach is a basic requirement and a necessity in today's world to harden the infrastructure. We close many exploits that could be open if we plan our baseline security well.

Vulnerabilities on devices such as ICS and SCADA need to be addressed and may require working closely with vendors to always be on top of patching. Zero-day vulnerabilities can pose a huge risk and need to be identified and mitigated immediately by looking out for indicators of compromise. Since we are talking about vulnerabilities, we can already imagine having focused patching for all enterprise-wide devices. The IoT, OT, and IT are all crucial, and taking a piecemeal approach to patching is a recipe for disaster.

Continuous monitoring is key for any security-focused organization. An organization focused on monitoring its IoT/OT assets one day that then lapses the next day will not really be effective at thwarting any attacks coming its way. With so many focused and agile attackers capable of even engaging in nation-state attacks, you need to ensure that your monitoring is always turned on and that alerts are sent to the SOC team in near real time as well.

This section provided more perspective on a holistic approach that you can take to securely manage IoT/OT devices.

Summary

In conclusion, there are multiple ways you can secure your organization holistically (IT, OT, the IoT, and the IIoT). In this chapter, you have seen the evolution and adoption of Industry 4.0 and how it has affected the cybersecurity landscape of IoT/OT devices. There are multiple approaches to adopting secure practices for the IoT or OT that organizations can take, and we have discussed some of the best approaches. To top it off, you can always look at defense in depth to ensure you have a backup security mechanism and are covered in the case of any failures.

In the next chapter, we will look at the network architecture of IoT/OT devices and how this affects cybersecurity when it is not managed well.

2

Delving into Network Segmentation-Based Reference Architecture – the Purdue Model

From introducing you to Industry 4.0, securing the OT/IoT infrastructure, the relevance of cybersecurity in the OT/IoT industry, and how to overcome some of the challenges present in OT/IoT, we'll move straight on to addressing the architecture followed in the industry today. It is very important to get a clear picture of the architecture implemented by organizations and the networks connecting devices.

In this chapter, we will cover the following topics:

- Zero-trust architecture
- Network segmentation in the IoT/OT environment
- Understanding the layers of the Purdue model
- How layers disrupt security when not managed well

Zero-trust architecture

It goes without saying that understanding zero-trust implementation is important to ensure the hardware root of trust is not breached and to ensure device integrity when we start architecting an organization's digital layout (including IT, OT, and IoT).

Zero-Trust Architecture (ZTA) is a strategy to gain the best possible security where nothing is trusted. It's a method for developing and putting into practice the following set of security principles:

- **Verify explicitly**: Always use the most up-to-date data points to authenticate and authorize.

- **Use least privilege**: Use data protection, risk-based adaptive rules, and **just-in-time** and **just enough access (JIT/JEA)** to restrict user access.

- **Assume a breach**: Minimize an attack's blast arc and divide the access area. Use analytics to drive threat detection, gain awareness, and strengthen defenses while verifying end-to-end encryption.

The zero-trust strategy helps organizations to be prepared for any adverse situation. It is in contrast to the situation in the past where every session that originated from a corporate network used to be treated as trusted and sessions from corporate devices outside the corporate network or coming from outside (i.e., the internet) were the only ones to be treated as insecure.

Organizations that have zero trust in place are as ready as they can be to handle failures and succeed.

Securing IoT/OT solutions with a zero-trust model is built upon the following requirements:

- **Maintain the asset inventory of all authorized devices**: Create a baseline of all the devices. Any unauthorized devices should be flagged immediately and be remediated with defined SLAs to avoid any security incidents.

- **Authenticate devices with strong identity decisions**: You may use the hardware root of trust that comes from a device's secure boot feature. Authenticate using passwordless or certificate-based authentication to make authentication and authorization decisions.

- **Mitigate the blast radius by maintaining least-privilege access**: Use access control for devices and workloads to reduce any potential damage from compromised identities, or those running unapproved workloads.

- **Continuous monitoring of device health for timely remediation**: Keep a watch on security configurations, do an assessment of vulnerabilities and weak passwords, and monitor for potential active threats and abnormal behavioral alerts to build risk profiles.

- **Regular updates to keep devices healthy**: Implement centralized configuration and compliance management solutions, as well as a robust update process, to ensure devices are up to date and healthy. Wherever updates are not supported by **Original Equipment Manufacture (OEM)**, ensure close monitoring by increasing the risk score and maintaining a high priority for the resolution of alerts.

- **Monitor direct internet access**: Any direct access on the devices in the OT network should be flagged immediately. Compensate with proper administrative control for allowing any internet access.

As described previously for OT/IoT security, an organization should consider built-in security for all upcoming IoT deployments. Consider an integrated approach to converge signals from legacy and modern protections. Identify the right device/service owners to take appropriate decisions to mitigate the risk and make contextual risk decisions to keep the business ahead of the competition and keep the process agile. With these ZTA concepts in mind, we will work toward understanding and implementing network segmentation in an IoT/OT environment.

Network segmentation in the IoT/OT environment

Network segmentation is an old but sure way of minimizing threats and protecting data or environments. This can be achieved both physically and logically.

With the arrival of Industry 4.0, industries are fast seeing the convergence of IT and OT. The advantages of bringing enterprise and industrial segments of networks together are simply too great to ignore. However, organizations need to take the utmost precaution when bridging the two historically isolated segments of the business as they bring together new vulnerabilities that are introduced by direct or indirect (through an intermediate device) internet connectivity.

OT systems were designed for a specific purpose—to last a long time and tolerate extreme weather conditions, in wet or caustic environments, and communicate over a proprietary protocol. These control systems were isolated from corporate networks. Hence, very often, we find OT networks are flat networks in these environments with hardly any security measures in place. This implies that any threat actor trying to penetrate the network will easily get through and may move laterally without any restriction, identifying critical assets or communications, and, therefore, discovering vulnerabilities and easily exploiting them.

A big, flat network in the OT environment is pretty common. This leads to a big threat if an adversary gets access to any resource, leaving the entire network vulnerable and prone to attacks. Network segmentation helps the organization to be better secured. Smaller/known segments help to apply controls in each segment, thus minimizing the impact in case of a breach. Monitoring and curtailing the attack helps protect against further spreading in case of an attack. *Figure 2.1* describes the purpose of network segmentation:

Control	Monitor	Protect
Network segmentation helps to control an attack within a network segment	Simplified monitoring to find anomalous activity faster in affected segment	Contain attacks within impacted network and protect from exploiting further network or enterprise assets

Figure 2.1 – Purpose of network segmentation

Perimeter security measures deployed to protect IT/OT environments are often insufficient as attacks do not always come from the outside. We may reduce the attack surface drastically by creating small segments to handle threats arising from outside or inside.

Network segmentation divides the computer network into smaller parts, thus enhancing network performance and security. Network segmentation is now evolved to significantly support a proactive network security practice in an OT/IoT segment. On one side, it segregates the system from the corporate network while still allowing the needed exchange of information; on the other side, it protects inherently insecure devices from direct exposure to the internet and other devices. Network segmentation helps in many ways, such as reducing the exposure of vulnerable devices, limiting cyberattack damage, and enhancing performance.

Given the benefits, network segmentation is good. Now, questions arise: can we use IT segmentation technology for OT/IoT segregation? The answer is *yes* and *no*. Yes, for segregating the corporate network from the OT network, but you may not use the same firewall to segregate networks within OT. The reason for this is that the firewall understands standard protocols such as HTTP, HTTPS, and FTP, while the communication within the OT network is based on the proprietary protocol, which the firewall may not inspect. Some next-generation firewalls may read some traffic but that's not enough to protect the OT network. Another point to note is you need to maintain the firewall with regular patches and upgrades, which may impact the OT network. Getting such frequent downtime in the OT network is almost impossible, and given the business impact, this is not a viable solution. It is similar for the network gateway. We need a solution that can cater to the requirements of an OT network with less maintenance and minimum effort. Data diodes are the mechanism prevalent in an OT system to meet the previously mentioned expectation of OT networks. We are going to talk about data diodes in a while as they help in creating one-way communication between OT devices and have the essential property of not requiring maintenance.

To overcome this, organizations started to map cybersecurity solutions to **Purdue Enterprise Reference Architecture** (**PERA**). The model helps in defining the segments in OT networks and the security policy for each zone. The zones are primarily defined based on the physical network architecture; however, the model falls short for business operations where it does not give clarity about maintenance, operators, vendors, and so on.

PERA, though, is not a reference architecture for OT cybersecurity but has been used to define segregation and control in industrial organizations. It helps greatly to define and plug security gaps in respective segments. It also helps IT and OT teams to comfortably talk about the issues in respective segments and remediation priority, thus bridging the gap in the control implementation. Step one is to establish the proper zones with clearly defined and enforced security policies. Step two is to properly secure the conduits with granular network traffic inspection. Step three is to create the baseline to flag anomalous behavior.

Another practical tip

Use people, processes, technology, and architecture together to solve the problem.

The takeaway: You need appropriate protection even within the network. Precision is the key to knowing what is flowing, where it is going, and how it is being used.

So far, we have understood the purpose of network segmentation in the IoT/OT industry and how it is used. We will now get an understanding of the layers of this network segmentation in the next section.

Understanding the layers of the Purdue model

You might have already obtained an idea of the Purdue model. Let's delve a little deeper into the layers it has and how it contributes to Industry 4.0.

Figure 2.2 gives us a clear representation of the Purdue model:

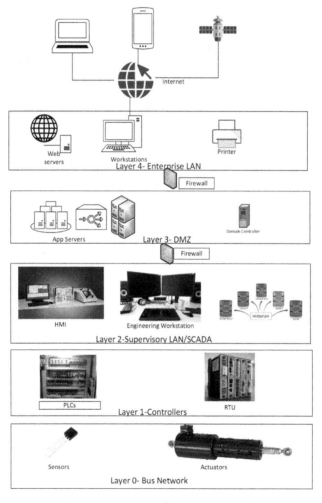

Figure 2.2 – Layers of the Purdue model

Let us look at these layers in detail:

- **Layer 0 – Bus Network**: This involves physical components such as sensors, actuators, and motor pump valves at the end of the cycle to produce the final product. Think of smart manufacturing—smart factories assembling products—as an example. These are called intelligent devices.

- **Layer 1 – Controllers**: The logic used to send commands to the devices at Layer 0 is from systems at Layer 1. Examples include **Programmable Logic Controllers (PLCs)**, **Remote Terminal Units (RTUs)**, **Supervisory Control and Data Acquisition (SCADA)**, and **Distributed Control Systems (DCSs)**. Control systems such as PLCs, DCSs, and RTUs monitor and control the physical devices at Layer 0, while SCADA also collects and sends data to the upper layer.

- **Layer 2 – Supervisory LAN**: The production workflow is taken care of at this stage. **Manufacturing Operations Management (MOMS)** or **manufacturing execution systems** are the systems that help in executing workflows such as batch management, recording data, and managing operations. **Human-Machine Interfaces (HMIs)** and engineering workstations that are connected to the PLCs/RTUs will further allow humans to control and monitor the physical devices. Data received from Layer 1 is stored in historians/databases.

- **Layer 3 – DMZ (Demilitarized Zone)**: This is the security system zone, which includes devices such as firewalls and proxies. This also forms a delineation for OT and IT. With the surge of automation, we can see that the data exchange between OT and IT has increased as well to aid in having an efficient environment. Thus, this IT-OT convergence has increased the adoption of Industry 4.0 as well, while ensuring that an attack can be thwarted with the right tools and devices.

- **Layer 4 – Enterprise LAN**: We are finally at the IT network/corporate network that we know today. Enterprise resource planning comes under this layer and it drives schedules, inventories, and so on. This is directly linked to business/revenue loss upon a disruption at this layer.

All these layers also represent the importance of segmentation. Also, in today's world, we have variations of the Purdue model as devices connect to the internet/cloud directly.

How layers disrupt security when not managed well

Grouping similar systems together to ensure we balance both performance and security is the key goal of the Purdue model. We have learned so far that segmentation and isolation are critical in ensuring security.

We are also clear on the benefits the convergence of IT and OT has brought to the industry. However, it has brought along with it attacks once only aimed at enterprises or IT now aimed toward OT assets as well. Every device is deemed mission critical as an attack on a front-line device or workstation or engineering system can lead to an attack on OT systems. This not only affects the business, revenue, and reputation but also human life, depending on the type of OT infrastructure.

Here is a scenario about the reconnaissance activity happening for TCP port 502. These activities not only have increased in the recent past but also, if successful, can issue harmful commands to OT devices.

TCP port 502 is commonly used in SCADA devices at Layer 2. This port uses Modbus, which is an app layer messaging protocol. This protocol can give access to physical devices connected to the internet.

Although we expect that the OT device is following segmentation and isolation as it is behind the firewall in DMZ, a successful attack on the internet-connected device is possible as the authentication standard used and the transmission of passwords in plain text can hamper security.

We are also well aware of the **Confidentiality, Integrity, and Availability (CIA)** triad and how important it is to IT security.

Here is a brief explanation of the CIA triad:

- Ensures data is kept confidential and accessed by the right audience
- All edits to the data are to be monitored and allowed/disallowed to make changes and keep the integrity intact
- It is also very important to ensure that the systems and corresponding data are always available and that there is no or minimal downtime

This changes slightly for the OT industry. Control of physical devices cannot be breached by any unknown entity or attack in an organization to always ensure *safe* operation. The concept of **Control, Availability, Integrity, and Confidentiality (CAIC)**, as represented in *Figure 2.3*, is significant:

IT | CIA   OT | CAIC

| Confidentiality Integrity Availability – Data/Apps | Control Availability Integrity Confidentiality – Physical Devices, such as Sensors, Actuators/OT Devices |

Figure 2.3 – IT versus OT priorities

Control of the device cannot be lost at any given instance as it could be a mission-critical device. Availability, for the same reason, is very important too and OT devices need to be available 24/7 as well. The integrity of the operational temperature or pressure changes for devices at Layer 0 can have a significant impact on safety as well. However, confidentiality for the OT industry isn't as important as it is for the IT industry.

With some of the scenarios mentioned previously, it is very clear how CIA and CAIC may be impacted if the layers are not protected well. However, if you look closely, this may impact overall security. When data/information travels between the IT and OT environments, if a threat actor gets access to the HMI or engineering workstation through direct/indirect internet connectivity, it is very easy to laterally move to IT and vice versa. A compromised workstation in the IT segment may allow a threat actor to gain easy access to OT and disrupt the business operation. The overall effect of this kind of attack would lead to the compromise of both CIA and CAIC, which may result in not only a compromise of CIA but also an impact on the business, which, in turn, may prove fatal for the plant operation team.

Adding a kill chain here (*Figure 2.4*), as it also applies to the OT network and clearly indicates when OT/IT/IoT isn't managed well, would result in the following:

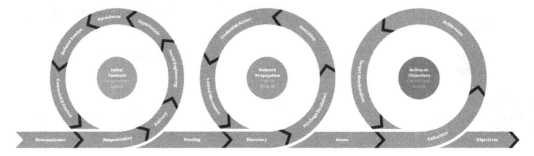

Figure 2.4 – Attack kill chain (Source: https://commons.wikimedia.
org/wiki/File:The_Unified_Kill_Chain.png)

Let us delve into the steps an attacker might take, as also seen in the attack kill chain, to compromise an IT/OT environment:

- **Reconnaissance**: Reconnaissance of initial access is possible at Layer 4 (Enterprise LAN) or directly at Layer 0 (Bus Network) as well if the device is directly connected to the internet with legacy security protocols.

- **Execution**: The attacker executes malware on the network or the device at various layers, finally making their way to the physical OT assets and thus ending up controlling the environment.

- **Persistence**: We can expect to see the attacker persist in the environment, say at Layer 4, and extract/collect as much information as possible about the environment to traverse further laterally.

- **Command and Control (C&C)**: Once a foothold is achieved, the attacker will circle back with the C&C, execute more commands, and follow through with their objectives.

- **Impair process**: The attacker ends up controlling the physical OT/IoT devices at the same time, also compromising quality and safety system functions.

A data diode is a solution to ensure that the network segmentation, when done well, raises the security bar.

Data diodes

A data diode is a physical device that allows only one-way data flow. It works on the principle that the diode basically acts like a DC switch. It allows current to flow easily in one direction but limits current flow in the opposite direction. This way, devices can send data and are immune to remote cyberattacks due to their physical nature. There are multiple usages of a data diode:

- **Safeguard one-way data flow and meet regulatory standards**: Data diodes are purpose-built devices with technology to move data only one way and enable fast and reliable data collection and sharing across different networks.

- **Secure file transfer and create a physical network separation**: Hardware-based diodes may enable segregation between the source and destination. This helps a regulated organization share the needed data in physically separated segments with the intended recipient without fear of any violations.

Data diodes in action in OT/IoT

The following are a few typical use case scenarios for traffic traveling to/from remote sensors and other facilities, replication of databases and other application data, and backup and disaster recovery repositories.

Create a physical border between the sender and destination networks to physically isolate them from one another while allowing for unidirectional data transmission. As seen in *Figure 2.5*, in order to protect networks from intrusions, malware, and other threats, this air gap between networks takes advantage of the principles of physics:

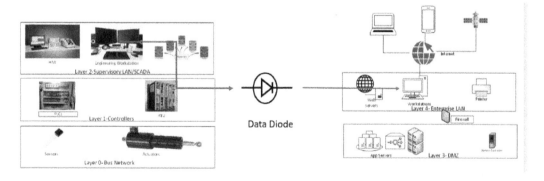

Figure 2.5 – Data diode enabling one-way communication between IT and OT

Data diodes are a great way to segregate networks. Being a device, they are not based on software, and as they cannot be controlled by code, they are immune to remote cyberattacks. The advancements in data diodes have helped organizations to create a one-way data flow with confidence.

Summary

In conclusion, we can say that ZTA is a long-term plan for many organizations following Industry 4.0. However, it should be started early with a vision to achieve the principles of ZTA in the long term. Doing so also ensures that network segmentation is achievable through the Purdue model or variations of it. Security at every network segment is critical as we have seen that a breach at any point can lead to lateral movement and compromise of the physical IT/OT/IoT assets, as explained in the kill chain. The security can further be enhanced by using data diodes and their one-way transfer functionality.

In the next chapter, we will learn more about attacks and threat vectors in the OT/IoT environment.

3
Common Attacks on IoT/OT Environments

OT is a critical network segment of businesses. OT deals with critical business processes and industrial operations. In a wide range of industries, such as the manufacturing, chemical, oil and gas, and power industries, it uses specialized technologies to run things such as assembly processes, production, floor operations, and energy grids. Over the years, processes in this segment have been automated by using OT, ICS, and SCADA systems, and fully automated plants have been attained in some industries, thus enabling agile, faster production.

All these OT operations were traditionally independent internal networks without any connectivity to the internet or corporate networks, and all OT operations used to be managed by an OT operations team. These industrial plants help businesses make millions of dollars worth of goods/services. Critical IT/OT infrastructure, which supplies water or power, helps the masses to get clean water and electricity. So, a simple downtime in such an automated plant may cause losses in the millions to a company and may put humans in the plant and the surrounding community at significant risk.

We are, in fact, now witnessing increasing attacks in the OT segment and this is causing real damage.

In this chapter, we will cover the following topics:

- Why do we see frequent attacks on the OT/IoT environment?
- Who performs attacks on OT/IoT systems and how and why do they do it?
- Famous IoT attacks
- How do these attacks impact businesses and humans?

Why do we see frequent attacks on the OT/IoT environment?

To unearth the truth regarding why IoT and OT are being attacked more now, let us understand the underlying cause that has left systems more vulnerable and attack-prone.

Diminishing airgap

A disconnected network is a type of network security that was leveraged in organizations in the past. This type of disconnected network is called an **airgapped network**.

An airgap was never a complete security solution, but network segmentation did pose a few challenges to the attacker.

OT networks were never completely airgapped. A vendor connecting their laptop to an OT network with a dongle connected to it for internet access. Remotely upgrading **Industrial Control Systems** (**ICS**) or remote firmware was all happening without knowing about these malpractices. IT/OT convergence increased the attack surface drastically and opened a new attack path.

Fully automated plants further encouraged businesses to utilize AI and ML to further boost the economy of operations, production, build-to-field time, and more by integrating external factors for more accurate prediction and supply. This introduced a new element to OT, *network connectivity to IT and the internet*. This led to another great attack vector that increased the OT attack surface drastically.

The legacy of OT assets

It is pretty easy to launch an attack on legacy devices with known vulnerabilities that are exploitable, making it quite cost-effective for hackers. Once the critical IT/OT infrastructure is compromised, it may yield higher payoffs, thus making it more economically viable with higher returns for the attackers.

As we can clearly see, we need to break the playbook of the attacker when it comes to IoT/OT critical infrastructure, and an understanding of the attacker—who they are and why they do it—will definitely help deter them.

Who performs attacks on OT/IoT systems and how and why do they do it?

The growing digitalization of organizations and IoT/OT convergence have brought new responsibilities for **chief information security officers (CISOs)/chief risk officers (CROs)**. They should address these and prepare for the risk imposed by internet-connected devices. Now, the question arises: why do we see frequent attacks on OT/IoT environments? This question can be answered in various ways, in terms of the intent of attackers, the known weaknesses, easy access, and the minimal cost of executing attacks. Before we venture to understand hackers' motivations, it is also particularly important to understand a little bit about cyber warfare.

Cyber warfare is defined as techniques, tactics, and procedures used during a cyber war by state-sponsored hackers. There are plenty of these techniques available. Some important ones are described here:

- **Espionage**: Spying or using spies to steal secrets from another country or business; this may be used by corporations to derive a business advantage by stealing competitors' ideas or trade secrets or by state governments to obtain political and military information on other countries. Some examples of this date back to 1712, when Europe obtained the details of China's porcelain-making process.

- **Sabotage**: Intentional and malicious acts that may cause the disruption of normal operations and functions or may cause the destruction and damage of IT systems or information.

  Stuxnet is a classic example of cyber sabotage that impacted critical infrastructure and is sometimes mentioned as the first widely known attack on critical infrastructure.

- **Denial of Service (DoS) attack**: An attack on websites or services with massive amounts of requests, bringing down the service and impacting legitimate users who are not able to use the service. These attacks may also include physical damage to worksites or network connectivity to a building, such as disconnecting the last mile of the cable in a building or cutting undersea cables.

- **Attack on national critical infrastructure**: This involves attacking the critical infrastructure of a country, such as power grids, water supplies, health and safety departments, the stock market, payment systems, and banks. This may disrupt the normal operations of the country and may have far-reaching effects. An attack on power grid safety instruments may cause damage to human life and infrastructure; it may cause blackouts and a prolonged restoration may have an impact on other dependent services, such as telecommunication and payment systems, thus heavily impacting finances.

- **Attack on the economy**: Most attacks directly or indirectly impact the economy of the target. However, targeting financial institutions such as banks, payment systems, stock markets, or other such services may jeopardize the economy of the country, where, for instance, citizens may not be able to use e-payment instruments to buy goods and services they need.

Figure 3.1 explains the different components of cyber warfare and how they are interlinked or overlap:

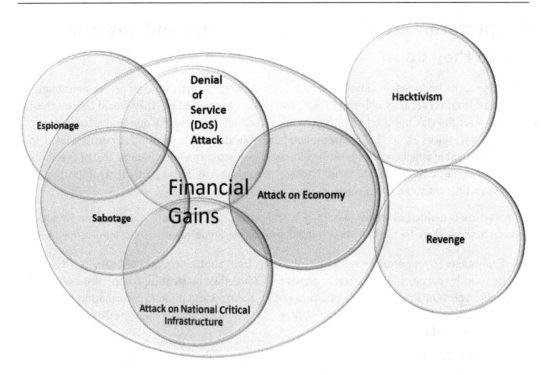

Figure 3.1 – Components of cyber warfare

Based on some famous attacks in OT/**industrial control systems (ICS)** environments, the following are some of the motivations behind cyber warfare:

- **Financial gain**: A cyberattack is most successful if hackers can get money from their sponsor or from victims in the form of ransom or compensation or by selling stolen data. This can be done in multiple ways, such as by installing ransomware. The Oldsmar water plant, for instance, paid vast amounts of money to a hacker to be back in operation quickly without any further business losses.

- **Revenge**: An internal employee, competing organization, or another country that is not happy and wants to punish the organization or a country to take revenge.

- **Hacktivism**: These actors are motivated by a philosophy and want to oppose and criticize a nation, government, or organization. For example, hacktivists could compromise a website to promote a political agenda and cause civil disorder.

- **Military**: A country's military may benefit from attacking its opponent's cybersystem and defeating them. This may result in the opponent giving up control without indulging in a physical war, which is much costlier.

- **Civil war**: Cyberattacks may also contribute to major disruption to a country's peace by instilling fear in the public or enticing them to revolt against the government or opponent.

- **Research/defense**: Sometimes, organizations that are attacked need to defend themselves and regain access to their infrastructure and they take measures to trap or manipulate the network. The sinkhole to prevent WannaCry is an example. Two UK-based cyber researchers registered a web DNS that they discovered during the WannaCry outbreak in May 2017. It is believed that WannaCry ransomware impacted more than 230,000 devices worldwide. The impact could have been more severe if this measure were not taken.

We spoke about the differences between cyber warfare and cyberattacks and the various motivations of an attacker. Let us see how these motivations have led many attackers to create havoc across industries in the next section.

Famous OT attacks

Cyberattacks on critical infrastructure can be very catastrophic and fatal too. Let us look at some attacks that have occurred in the recent past and how they have affected the industry and the people involved.

The Triton attack

When we recount critical infrastructure attacks in the recent past, we cannot miss the Triton attack, which happened in 2017. This nation-state attack on ICS happened in the Middle East on Schneider's Triconex safety systems in a petrochemical plant. This attack on a safety instrument system with malicious code that could eventually lead to the release of toxic gas was one of a kind and could have been fatal to human life.

Now, how did the attacker gain access to the network in the first place to have executed this remote code? You will not be surprised to find out that the action started with a spear-phishing campaign.

Figure 3.2 illustrates this attack in a diagram for a better understanding of the steps that could have been followed by the attacker:

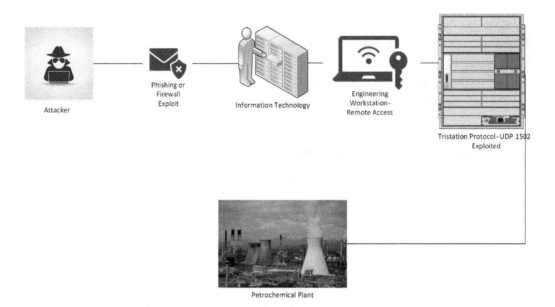

Figure 3.2 – Flow of attack in the Triton attack

An attacker successfully reaches an engineering workstation and compromises it further with a phishing email, blurs the firewall between IT and OT, further executes a TriStation protocol through UDP port `1502` on critical controllers, and completely compromises the system, leading to the shutdown of the plant. The result could have been far worse or fatal, but was not, and thus led to the assumption that the final payload wasn't executed.

Conclusion: Do not shy away from security audits, especially if you operate systems that would lead to devastating/fatal results if attacked. Firmware updates are also key to ensuring less exploitation.

Oldsmar cyberattack on the US water system

The water treatment system of Oldsmar, Florida, was attacked and the attacker gained access to the process control of the treatment system on February 5, 2021.

The attacker was successful at increasing the sodium hydroxide content in the water systems to toxic levels. Talk about OT systems affecting human life when manhandled! Thankfully, in this case, the operator was alert and immediately reduced the toxin levels in the water system and stopped the entire city from being poisoned.

Figure 3.3 illustrates this attack in a diagram for a better understanding of the steps likely followed by an attacker:

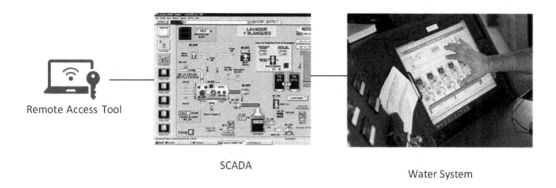

Remote Access Tool

SCADA

Water System

Figure 3.3 – Flow of attack in the Oldsmar cyberattack

Understanding the systems used at Oldsmar can give us more insight into what caused the attack. They used SCADA, which would in turn help them monitor and control the facility. They were said to be using Windows 7 and had no real firewall protection as well. To top it all off, the TeamViewer app they used had the same password shared among many employees.

Conclusion: Real-time monitoring is key. It is necessary to update devices to support an operating system for which the security patches are still being released (e.g., Windows 7 is out of support and patches aren't released for it anymore, hence it is advised to upgrade to a supported OS such as Windows 10 or 11). Sharing passwords should be avoided and MFA should be used to access systems. We cannot forget the core security components, such as antivirus, firewalls, and patches. Be wary of using remote collaboration tools. Use them, if necessary, with a centralized authentication system and with MFA.

The Colonial Pipeline cyberattack

Colonial Pipeline carries jet fuel and gasoline in the US region. The ransomware attack, which is deemed to be the largest on oil infrastructure in the US, occurred on May 7, 2021, and the infrastructure was down for nearly 6 days before service restarted on May 13, 2021. This successful ransomware attack was on the billing infrastructure of the company. However, this brought down the entire oil pipeline as they could not issue customers with appropriate bills. This oil line attack affected the entire east coast of the US.

Figure 3.4 illustrates this attack in a diagram for a better understanding of the steps followed by the attacker:

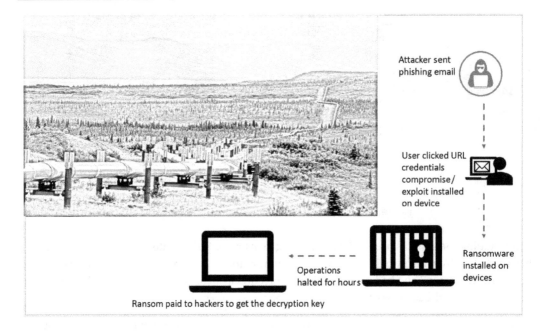

Figure 3.4 – Flow of attack in the Colonial Pipeline cyberattack

Here, the attacker compromised the user system with a phishing email and managed to successfully install an exploit. This led to the encryption of devices as the ransomware spread laterally in the organization. This halted operations for hours, affecting many in the Texas region. The company decided to pay the ransom. This, however, is not a recommended solution as it will encourage the ransomware attackers further and you cannot be sure that the attacker will provide you with the decryption key. Not getting attacked in the first place and not letting it spread across the estate is a better approach to managing this threat vector.

Conclusion: Phishing emails need to be filtered out. Users need to be educated to recognize phishing emails. Endpoints need to have monitoring setups to immediately identify, remediate, and, if necessary, automatically isolate the device from the network upon detection of malware to prevent it from spreading across the network.

The Ukraine electric grid attack

Although this attack occurred on December 23, 2015, it still serves as the biggest example of a cyber-attack on OT systems and can teach us multiple things.

A spear-phishing email using black energy malware was sent to an employee. When the employee was compromised by the phishing email, the attacker got access to the network. The attacker stayed in the system, moving laterally to the OT network from the engineering workstation, and then compromised SCADA. Once SCADA was compromised, the attacker persisted laterally, looking for vulnerable hosts and installing malware further.

Finally, the HMI was attacked, and the threat was activated, turning off the electric grid or breakers. This is depicted in *Figure 3.5*:

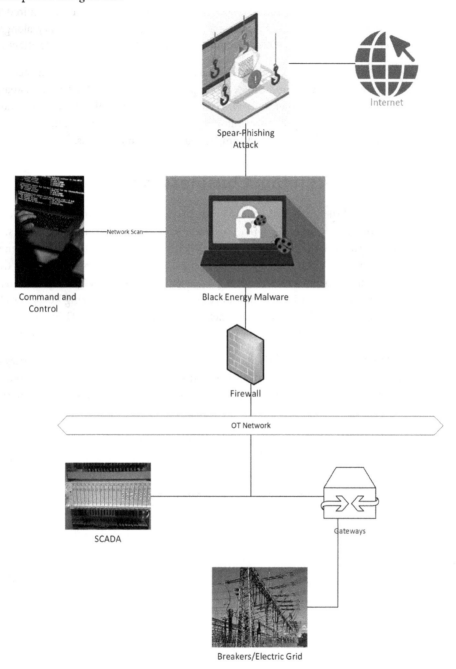

Figure 3.5 – Flow of attack in the Ukraine electric grid attack

Conclusion: Protection against phishing is a must. IT protection is still the first step. The endpoint is to be monitored for any attacks and to stop executions of scripts in real time and to stop connections to the command and control center. Better firewall protection, as explained previously, is required too. Real-time monitoring of the OT network and devices for any potential change is necessary, alongside having the SOC team take immediate action to prohibit or thwart such large nation-state attacks.

Cyberattackers have shown they are still eager to attack vital infrastructure with hostile online behavior. Organizations should develop a precise and thorough map of their IoT/OT infrastructure, analyze and identify the specific risk related to their current OT devices using the validated asset inventory, and put in place an ongoing, vigilant system-monitoring program with anomaly detection. In the upcoming section, we will learn how these attacks impact businesses and humans.

How do these attacks impact businesses and humans?

We have, so far, seen some of the impactful attacks in the OT world and have managed to pick up some lessons from each of them. Most of them point us in the same direction: toward finding a stable solution to be able to apply to both the IT and OT worlds. We have seen some of the amazing benefits of the convergence of IT and OT with Industry 4.0 and how it impacts businesses, thrusting them forward toward success in the data and AI worlds. However, when the fast-paced digital transformation is not intertwined with the right security and monitoring, we will see it hampering businesses in a negative way. We have also learned from previous attacks that the impact on the OT world is visible not only on the business (monetary or reputation loss) but also, unfortunately, on human lives too.

It is especially important to understand the risk appetite of any organization. Based on this, they can focus on how to improve their IT/OT security posture. But when there is a missing link or one too many missing links, such as the following points, business stakes become extremely high and losses are tough to measure as well:

- **No asset discovery**: IT assets, data, databases, security cameras, sensors, smart factory assets, and so on, as shown in *Figure 3.6*:

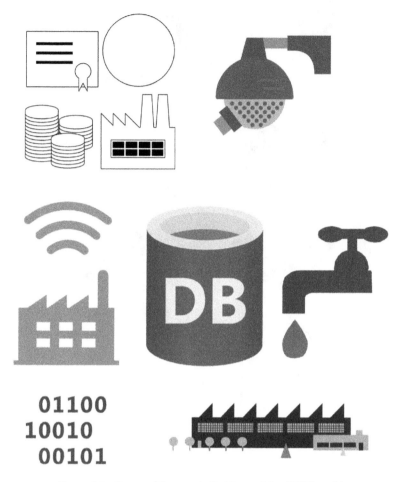

Figure 3.6 – Some of the assets that impact the IT/OT world

Assets can be unique to each organization, but to most organizations, they are devices, data, anything valuable, or any device paving the way to valuable information. Sometimes, organizations fail to identify all devices in the organization. This is an open security flaw as an unknown device within the network with no security can be an easy opening for any attacker.

- **Having no acceptable recovery time**: This is where a risk appetite would help an organization define how long an asset can go down without it affecting the business. The CIA Triad (depicted in *Figure 3.7*) explains that for any business, it is absolutely essential to ensure confidentiality, integrity, and availability of organizational resources. We learned previously that for the OT industry, integrity and availability are far more important. When an organization is able to put this (integrity and availability) together clearly, they will be able to come up with an acceptable recovery time for critical resources as well:

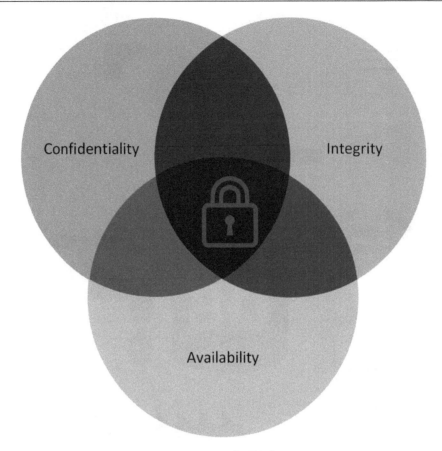

Figure 3.7 – CIA Triad

- **Miscellaneous**: Along with a bad firewall policy, no systematic patching, not upgrading to current firmware requirements, no measures to stop phishing emails, and no real-time monitoring on endpoints for observing and stopping attacks in real time can have a huge business impact, as well as impacting human lives.

Hence, it is important to have a strong security approach; ideally, one that aligns your IT/OT infrastructure with the **Zero-Trust (ZT)** framework, by understanding the risk appetite of your organization.

Summary

OT attacks are critical and have leaned more toward nation-state attacks; hence, it is crucial to have an overall security system, a zero-trust approach, and a defined risk mitigation approach after finding a versatile asset discovery approach. In the next chapters, we will learn how to implement all these (security, a ZT approach, risk mitigation techniques, asset discovery, and so on), especially focused on OT/IoT security with Microsoft Defender for IoT.

Part 2: How Microsoft Defender for IoT Can Address the Open Challenges in the Connected World We Live in Today

In this section, the focus is on learning about Microsoft Defender for IoT and how its features help with securing OT-IoT environments. We will also learn about Microsoft Defender for IoT's architecture and its deployment.

This section includes the following chapters:

4

What Is Microsoft Defender for IoT?

Microsoft Defender for IoT (**MDIoT**) empowers IT and OT teams to identify critical vulnerabilities and detect threats using behavioral analytics along with **Machine Learning** (**ML**) that is IT/OT aware. All this happens while also ensuring that performance and availability are not compromised.

It provides a centralized interface to manage the threats and vulnerabilities affecting the ever-increasing attack surface in the IoT/OT world.

MDIoT also provides network-layer monitoring that is agentless and can integrate seamlessly with industrial equipment from various vendors and SOC tools.

In this chapter, we will cover the following topics:

- The IoT and OT environments
- The role of asset inventory
- Risk and vulnerability management
- Continuous threat monitoring
- Operational efficiency
- MDIoT benefits

The IoT and OT environments

Figure 4.1 is a diagrammatic representation of IoT and OT environments in organizations, which might then be connected to an IoT hub for the further processing of data. Azure IoT Hub receives data from both OT and **Enterprise IoT** (**EIoT**) environments, as depicted in the following figure:

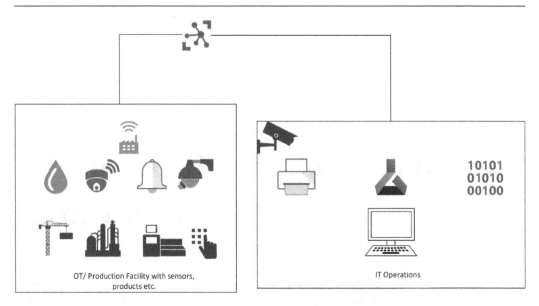

Figure 4.1 – Representation of IoT and OT environments

Let us look into **agentless device inventory**. The issue that the industry is trying to solve is keeping OT/IoT devices up to date, but in reality, they are unpatched, and IT teams have no insight into the status of the devices, whether they are patched or not. Additionally, these OT/IoT devices do not support agents. It is, hence, important to understand specialized protocols used by IoT/OT devices and machine-to-machine behaviors to be able to use agentless monitoring using MDIoT. Thus, it provides security and visibility into the organization's network.

A centralized portal can give a visual representation of IT, OT, and IoT devices to the security team so they can do the following:

- Discover assets
- Assess risks
- Manage vulnerabilities
- Carry out continuous threat monitoring
- Maintain operational efficiency

The role of asset inventory

Asset discovery, that is, finding the assets on your network, is the first step to achieving security. You can only protect and secure your known assets. OT environments generally lack an automated continuous asset inventory function. Based on our experience in the field, we have found that most of these inventories are in a physical register or spreadsheets, which is inventory taken when the device was onboarded and does not help much if you want to assess which devices in your environment you need to protect. *Figure 4.2* highlights the challenges involved in asset inventory for organizations. You can see that the manual method of asset inventory using Excel or pen-and-paper is time consuming and not very productive, and we would like to introduce a more effective means of asset inventory:

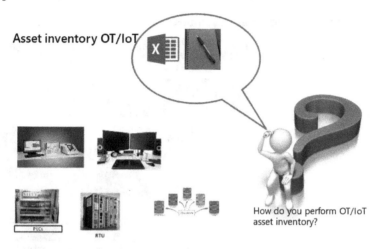

Figure 4.2 – Asset inventory challenges

In the IoT/OT industry, we need to identify all the assets that belong to any organization. This is a critical step as any asset missed can easily become an entry point for an attacker. *Figure 4.3* depicts assets in the IoT/OT industry being successfully inventoried in a centralized location. This can further be streamlined to allow SOC teams to view them in a SIEM tool as well:

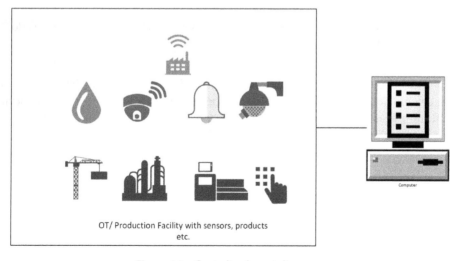

Figure 4.3 – Centralized asset discovery

MDIoT allows active and passive scanning of the network. The following diagram, *Figure 4.4*, depicts a combination of active and passive scanning. In order to get complete visibility, you may want to leverage Microsoft Defender for Endpoint as an active agent to monitor and protect endpoints such as servers, laptops, and tablets. This integrates very well with MDIoT to give complete visibility:

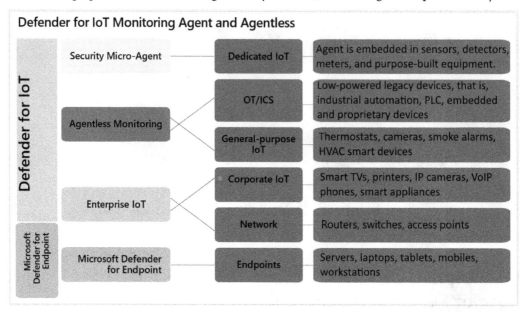

Figure 4.4 – Active-passive scan using a combination of MDIoT and Microsoft Defender for Endpoint

> **Note**
>
> You may choose to opt for a complete passive inventory, which is also known as a zero OT/IoT performance impact network scan.

Asset inventories help you see the asset list by adding and editing columns; sorting the devices by clicking the column header; filtering the list based on asset type, that is, all programming devices; exporting an entire asset list or filtered list in CSV format; exploring details of the device by selecting a device and viewing the full details; and removing the device from the inventory.

Asset inventories also help in visualizing IoT/OT network topologies and analyzing heterogeneous and proprietary industrial protocols to find established communication paths, and further use this information to hasten network segmentation and the zero-trust journey:

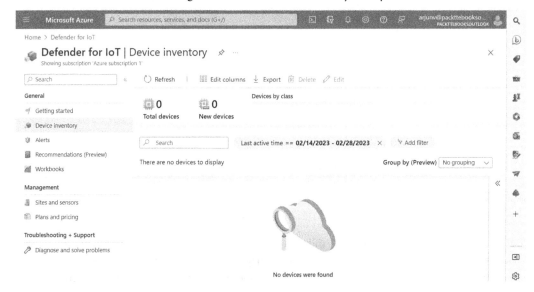

Figure 4.5 – Screenshot of asset discovery from the MDIoT portal

In this section, we have explored the importance of asset inventory in the IoT/OT industry, and have given a glimpse into how this can be achieved via MDIoT. This will be elaborated upon further in *Chapter 7, Asset Inventory*.

Risk and vulnerability management

Awareness of the risks and vulnerabilities present in customers' OT/ICS networks may help them to plan and mitigate the associated risk.

An OT network with many assets and sensors that are successfully tracked and inventoried in the asset discovery needs to be analyzed for any vulnerabilities. Ideally, any vulnerabilities also need to be remediated before they are exploited by an attacker. The block diagram in *Figure 4.6* depicts assets' vulnerabilities being remediated:

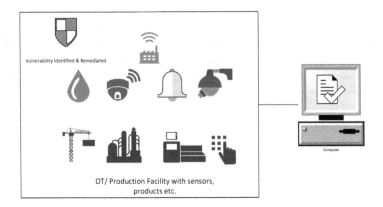

Figure 4.6 – Vulnerability management for assets in the IoT/OT industry

Some of the techniques used by MDIoT for learning vulnerabilities include **Deep Packet Inspection (DPI)**, **Threat Intelligence (TI)**, **Artificial Intelligence (AI)**, and ML.

MDIoT helps the IT team identify unpatched assets, device configuration issues, unauthorized applications, top attack vectors, open ports, unauthorized connections, and missing anti-virus and then prioritizes mitigation based on asset importance and direct exposure. We will go through the details in *Chapter 9, Vulnerability Management and Threat Monitoring*. In *Figure 4.7*, we can see a snippet of the risk assessment report. The executive summary provides visibility of the devices and anomalies and some recommendations to fix any risks as well:

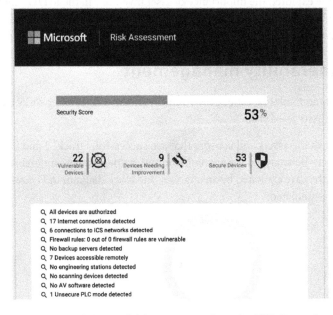

Figure 4.7 – Snippet of risk assessment from the MDIoT portal

Attack vector reports show a visual representation of a chain of vulnerable devices' vulnerabilities. An attacker may be able to access important network equipment because of these vulnerabilities. The **Attack Vector Simulator** examines every attack vector for a given device and calculates those attack vectors in real time. *Figure 4.8* depicts how an attacker could navigate a network to compromise the device/asset:

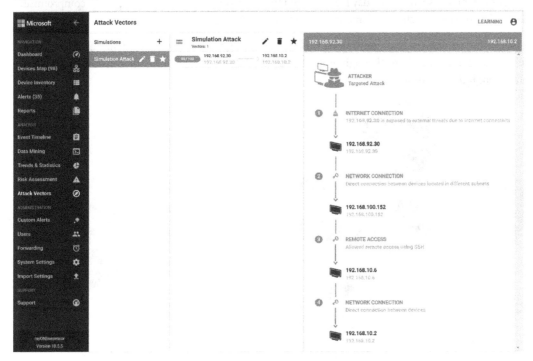

Figure 4.8 – Sample attack simulation report

In this section, we learned how asset inventory enables a SOC team to find devices' risk status and bring visibility to all possible attack paths for the device in question.

Continuous threat monitoring

Microsoft's **Section 52**, the MDIoT security research group, is a team of passionate OT threat researchers, nation-state defenders, and data scientists. The team does OT/IoT threat hunting, malware reverse engineering, protocol search, and OT cyber-incident response. The information provided by **threat intelligence** (**TI**) feeds helps in identifying threats in the IoT/OT industry and thus aiding in stopping adversaries from exploiting vulnerabilities. The TI is pushed to the MDIoT cloud-connected sensors at regular intervals; offline sensors need to be updated at a regular frequency.

Recognizing targeted attacks and malware by leveraging threat-hunting tools and behavioral-aware analytics by scanning through historical network traffic and **Packet Captures** (**PCAPs**) is done continuously. This is a key feature of MDIoT as it will flag an alert if an attacker is trying to make unauthorized changes in endpoints, assets, or sensors. So, we (the SOC team) are there to intervene before an attacker is able to shut down an entire electrical grid or any other critical safety system, which could pose a threat to human life if allowed to proceed.

Modern-day threats, such as zero-day malware, and the tactics missed by static indicators of compromise can be identified with behavioral analytics provided by MDIoT.

In *Figure 4.9*, we can see that continuous threat monitoring leads to cyber resilience as it leverages the asset inventory and skims through multiple real-time alerts, PCAPs, and other historical network data through threat-hunting tools:

Figure 4.9 – Continuous threat monitoring

In this section, we explained how continuous threat monitoring may help organizations to be more cyber resilient.

Operational efficiency

MDIoT is built from the ground up for OT security. It sends real-time alerts about device misconfigurations and malfunctions. This is an additional capability on top of the cybersecurity benefits. This capability helps the OT operations team to identify and carry out root cause analysis by flagging abnormal activities and providing full details that are needed from the OT segment and for any operational issues that may impact plant availability or product quality.

In *Figure 4.10*, an operational alert is highlighted to show that MDIoT is capable of recognizing misconfigurations. This functionality aids in increasing operational efficiency as admins are able to employ the necessary remediation steps:

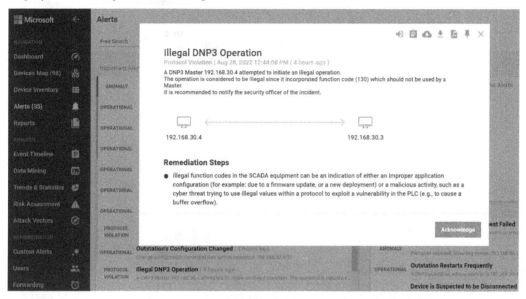

Figure 4.10 – Screenshot of an operational alert aiding in operational efficiency

We will get into the details regarding understanding the alerts in the coming chapters.

At the end of this section, we now clearly understand the capabilities of MDIoT. In the next section, you will see how these capabilities, when put together, can benefit any organization in the IoT/OT space.

MDIoT benefits

MDIoT leverages passive monitoring and **Network Traffic Analysis** (**NTA**). This is the highlight of the product as it combines passive monitoring and NTA with Microsoft's own patented technology (i.e., IoT/OT-based behavioral analytics) to capture information in real time.

To capture network traffic, we need to deploy a sensor on-premises to a network SPAN port, and we shall now see some of the advantages of this.

Zero impact on network performance

Most OT environments cannot sustain an active scan. *Figure 4.11* represents the fears of a plant operating team about downtime caused by an active scan:

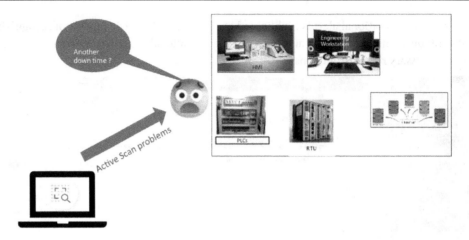

Figure 4.11 – Active scanning OT devices may cause downtime

Downtime directly translates to a loss of business or profit, which is a central cause of worry for most organizations. MDIoT is here to help as it leverages passive scanning of network traffic. The huge advantage of this is that the device being scanned will not be actively impacted. Network-scanning tools such as **Nmap** and **Nessus** have the capacity to actively ping the devices and at times bring them down too.

This out-of-band inspection of network traffic has zero performance impact on the OT industry/ organization using it.

Figure 4.12 represents passive monitoring being carried out by the MDIoT sensor through the SPAN port:

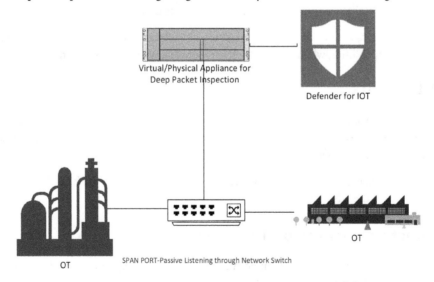

Figure 4.12 – Passive monitoring of network traffic through the SPAN port

The devices aren't affected as network traffic is analyzed passively. This is thus the best fit for locations with low bandwidth and high latency.

Quick deployment

You'd be surprised to see that the behavior-aware sensor deployed in the network starts generating insights within minutes of deployment. The archaic method of setting rules or even signatures can be left behind, as this solution leverages ML to set up rules automatically by recording the environment and determining what is normal. We will expand on this topic in the next chapter, *Chapter 5, How Does Microsoft Defender for IoT Fit into Your OT/IoT Environment/Architecture?*.

Advanced threat detection

MDIoT detects sophisticated IoT/OT threats, such as **living-off-the-land** tactics and fileless malware, on the grounds of unauthorized or unusual behavior. MDIoT uses a patented approach combining the layer-7 DPI method with **Finite State Machine** (**FSM**) modeling, which views the behavior of IoT/OT networks as a predictable series of states and transitions. As a result, MDIoT can identify threats more quickly and accurately after only a very small learning window.

> **Note**
>
> If you are an IoT device builder and are interested in MDIoT, there is also a lightweight micro-agent available supporting Linux and **Real-Time Operating Systems** (**RTOS**). This way, security is built into the device and your projects.

Summary

In this chapter, we have successfully introduced MDIoT and delved into its capabilities and features. Some of the highlights are asset discovery, risk and vulnerability management, continuous threat monitoring, and operational efficiency. We have seen its benefits, including passive scanning, quick deployment, and advanced threat detection. We will look at many of these in the next few chapters, so get ready! In the very next chapter, we will understand how MDIoT addresses the security gaps through risk assessment, continuous threat monitoring, and more, along with the steps required to deploy MDIoT.

How Does Microsoft Defender for IoT Fit into Your OT/IoT Environment/Architecture?

Understanding **Microsoft Defender for IoT** (**MDIoT**) and its features has been a good start so far. Now, let us delve deeper into understanding how MDIoT can fit into an organization's network or architecture.

In this chapter, we are going to learn about network sensor placements in various network topologies. We will also learn about Azure Cloud-connected sensors, the Azure portal, and an on-premises management system to aggregate and manage multiple network sensors from a single system.

This chapter will cover the following topics:

- The topology of network architecture
- Diverse ways of traffic mirroring for OT monitoring
- How the Purdue model is applied to MDIoT
- Sensor placement considerations
- OT sensor cloud connection methods

The topology of network architecture

When planning your network monitoring, you must understand your network architecture and how you will need to connect it to MDIoT. It is also important to understand how each of your system elements falls into the Purdue model for **industrial control system** (**ICS**) OT network segmentation.

MDIoT network sensors continuously detect and monitor network traffic from IoT and OT devices. To get full coverage for your OT/IoT devices, the network sensor needs to be placed in such a way that it can read all the network traffic. Let us delve into this now.

The most common network topologies in OT/IoT networks

The following are some samples of MDIoT sensor placements for different network topologies:

- **Ring topology**: In a ring topology network, nodes or switches create a circular data path. Each switch or node device is connected to two others in a circular format. Collectively, the switches or nodes in a ring topology are called **ring networks**. Only one switch in the ring needs to be monitored. *Figure 5.1* depicts this topology:

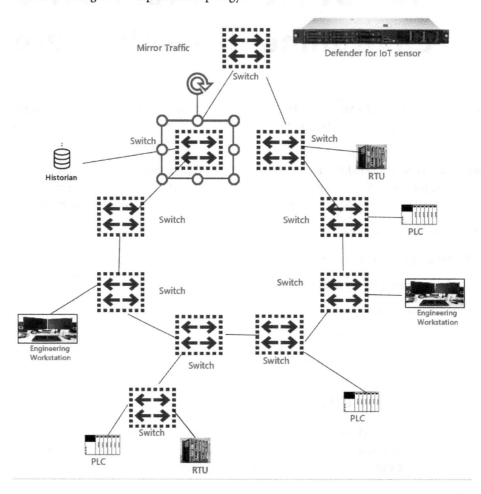

Figure 5.1 – The placement of an MDIoT sensor in a ring topology

- **Star topology**: In a star network, each switch or node is connected to a central hub. In simple terms, a central hub acts as a conduit for messages. In *Figure 5.2*, lower-level switches are not monitored by MDIoT, and the traffic that remains local to those switches is not visible. Devices can be identified using **Address Resolution Protocol** (**ARP**) messages, but the connection information cannot be read:

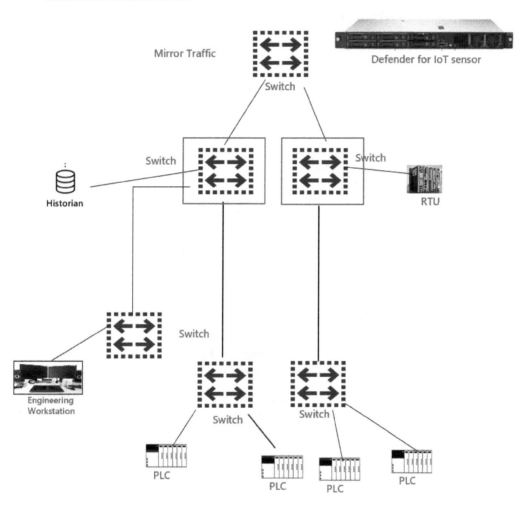

Figure 5.2 – The placement of an MDIoT sensor in a star topology

We have seen sensor placements in single-tenant setups so far. Now, let us understand sensor placements in a multitenant setup. This will be much more aligned with the real, complex setup of networks for many organizations as well.

A multilayer, multitenant network

IT service management, SIEM solutions, asset management solutions, and Azure services may all need access to the various layers of the Purdue model. This can be achieved via a multilayer, multitenant network, as shown in *Figure 5.3*:

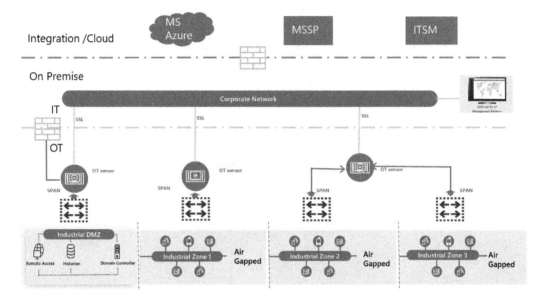

Figure 5.3 – Representation of a multilayer, multitenant network

Layers 0 to 3 of the OSI model usually have **network traffic analysis (NTA)** sensors. The data collected here can then be leveraged by or sent to the **managed security service provider (MSSP)** or the SOC team.

This section is intended to help you understand and decide on the network sensor placements for various networks. By now, you may understand your own network's OT architecture better and therefore, be ready to plan out your deployment. In the coming section, we will learn about methods for traffic mirroring and passive or active monitoring.

Diverse ways of traffic mirroring for OT monitoring

Network sensors, which are a significant part of the MDIoT architecture, receive data or traffic from the following:

- SPAN ports

- The network **terminal access point (TAP)**

In *Figure 5.4*, we can see that the OT devices send traffic for analysis through the managed switch with port mirroring:

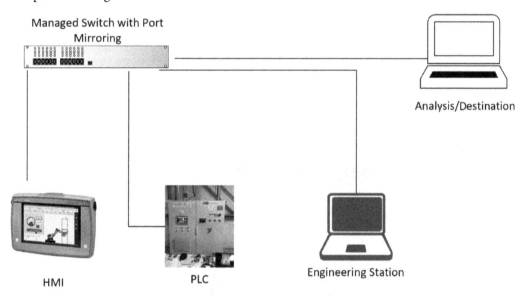

Figure 5.4 – Example of the network used in an OT environment

We will now learn about different methods used for traffic mirroring in an OT environment to enable monitoring with MDIoT.

To focus on specific and relevant network traffic for traffic analysis, you need to connect MDIoT to a network mirroring port on a switch or a TAP that only covers industrial ICS and SCADA traffic.

SPAN

Port mirroring, commonly known as **Switched Port Analyzer** (**SPAN**), is a method of monitoring network traffic. When port mirroring is enabled, the switch sends a copy of all the network packets visible on a single port (or across an entire VLAN) to another port, where the packets can be analyzed. This can be configured as follows:

1. **Local SPAN**: This directs traffic from one or more interfaces on a switch to one or more interfaces on the same switch.

   In *Figure 5.5*, we can see the mirrored managed switch port and that all nodes have a copy – everything on the network coming its way is mirrored for further analysis:

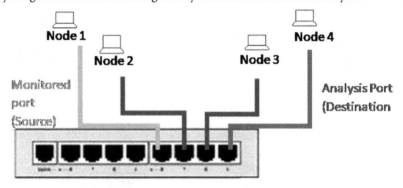

Figure 5.5 – Mirrored managed switch port

2. **Remote SPAN** (**RSPAN**): This allows you to monitor source port network traffic distributed over multiple switches and helps centralize the network sensor in the network. RSPAN works slightly differently than local SPAN, as it copies the monitored traffic onto a specific VLAN for the RSPAN session. The VLAN is trunked to all other switches in that network, which allows the RSPAN session traffic to be sent to multiple switches. The central switch configured as the destination port just mirrors the traffic from the specified VLAN to the assigned ports.

   In *Figure 5.6*, the network topology of RSPAN is depicted and we can infer that it copies the traffic and channels it into specific VLANs:

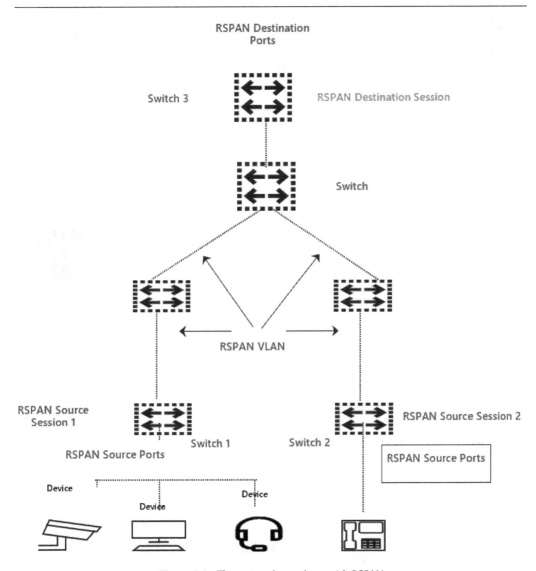

Figure 5.6 – The network topology with RSPAN

3. **Encapsulated Remote SPAN (ERSPAN):** This adds a **generic routing encapsulation (GRE)** layer for all captured traffic and allows it to be extended across Layer 3 domains and routed to other networks.

ERSPAN is a Cisco proprietary feature and is available on several family products, such as the **Aggregation Services Router (ASR)**, Nexus, and the **Cloud Services Router (CSR)**. *Figure 5.7* shows mirrored traffic being transported through an IP network. Here is how it is done:

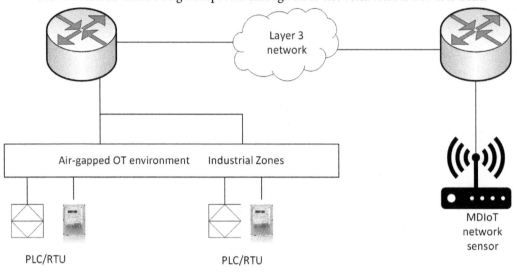

Figure 5.7 – ERSPAN

The source router sends the traffic by encapsulating it over the network. The packet is received at the destination router, which decapsulates it.

Active and passive aggregation

MDIoT sensors can be used with an active or passive aggregation TAP. *Figure 5.8* shows how it is installed right in between the network cable:

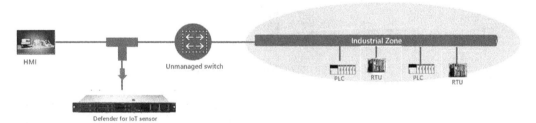

Figure 5.8 – Architecture of a network TAP

This device duplicates bidirectional (RX-Receive and TX-Transmit) traffic to the monitoring sensor. Industrial ICS and SCADA traffic will be captured and further analyzed for any deflections or alerts.

A TAP is a hardware device that efficiently copies this bidirectional traffic and ensures the integrity of the network at the same time. The biggest advantage of a hardware device is that it cannot be compromised as easily as a software device. Hence, the usage of TAPs is suggested in critical infrastructures – and for forensic purposes, too.

Another reason why you should consider TAPs is that they do not drop traffic, even if the messages are broken when a switch drops these packets. TAPs are not process-sensitive and the packet times are exact.

Some of the common types of TAPs are as follows:

- Garland P1GCCAS
- Ixia TPA2-CU3
- **US Robotics (USR)** 4503

We have reached the end of this section. You should have gained a lot of enriching information on the diverse architectures or network topologies possible for monitoring with MDIoT and the various devices/options you have to monitor the traffic. Armed with this data, let us further understand the Purdue model and how it is associated with MDIoT.

How the Purdue model is applied to MDIoT

So far, we have delved into understanding network architectures and their topologies. Now, we will get into details of how the Purdue model (representing the OT network segmentation) in your network architecture links to MDIoT.

Let us learn about the Purdue layers that we covered in the previous chapter, *Chapter 4, What Is Microsoft Defender for IoT?*, in the context of MDIoT:

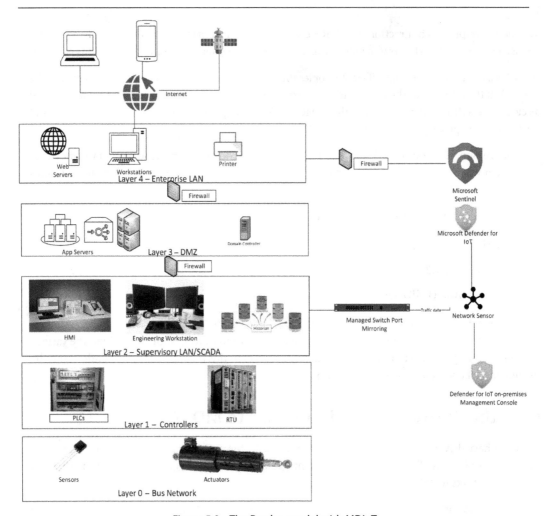

Figure 5.9 – The Purdue model with MDIoT

As we can see in *Figure 5.9*, starting from **Layer 0**, the layers of the Purdue model pass on information to the layers above them. **Layer 0** consists of sensors and actuators, which interact with the physical environment where necessary and pass on the data to the layers above. **Layer 1** consists of PLCs and RTUs. The data received from the sensors is further processed in these logical units and they provide adjusted output. **Layer 1** also focuses on connecting the hardware of actuators to **Layer 2**.

Layer 2, which holds our managed switch port, manages to collect the traffic from these devices (in **Layer 0** and **Layer 1**) and send it across to the MDIoT on-premises or online portal through a SPAN port.

Layer 3 (and sometimes, **Layer 2**) contains historians and an operation management system, which facilitates better analysis of the context and also stores the data. **Layer 3** can also host the AD server and file server in the DMZ environment.

Meanwhile, **Layer 4**, which is part of the IT environment, connects to the SIEM solution (in this figure, **Microsoft Sentinel** is the SIEM solution). Alerts, warnings, and messages are all sent to the online MDIoT console and on-premises MDIoT console and can be forwarded to the SIEM console. The alerts sent to the SIEM portal will keep the SOC team updated on any changes to the OT/IoT network.

Here, I am reiterating how important the Purdue model is and how MDIoT can depict all your OT/IoT devices in this (Purdue) format on your MDIoT portal. *Figure 5.10* depicts the clear delineation between the OT environment and the enterprise IT environment:

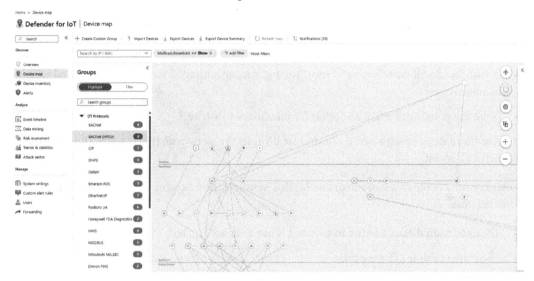

Figure 5.10 – The Purdue model represented in the MDIoT portal

As we traverse through further chapters, we will learn a lot more about the MDIoT portal and how this Purdue model-based depiction supplied by the various vendors used in the organization will make the lives of administrators easier.

Sensor placement considerations

Once you understand the target OT network architecture and how the Purdue model applies to it, you may start planning sensor connections in an MDIoT deployment.

Generally, MDIoT is used to monitor traffic from Purdue layers 1 and 2. However, in most modern organizations, OT traffic exists on layer 3 as well; therefore, you can use MDIoT to monitor layer 3 traffic.

Review your OT and ICS network diagram with site engineers to determine the best place to connect to MDIoT to get the most relevant traffic for monitoring. We encourage you to meet with local network and operational teams to clarify your and their expectations. It is a promising idea to create a list of the following information for the target network:

- A list of devices.
- The number of OT networks in the target site.
- The number of devices in the OT network.
- The vendors and industrial protocols in the OT segment.
- Network engineering managers and supporting external resources (if applicable).
- The number of switches and switch models and their capabilities to support port mirroring.
- If you decide all switches need monitoring, can unmanaged switches be replaced with managed ones?
- Is the use of network TAPs an option for unmanaged switches?
- Are there devices with serial communication in the network? If yes, mark them on the network layout.

You may consider deploying more than one MDIoT sensor if the following conditions apply to your OT/IoT network:

- The maximum distance between switches is more than 80 meters
- There are multiple OT networks
- The number of switches is more than 8, and the local SPAN port close to the sensor is 80 meters away by cabling distance

The best part here is that the MDIoT sensor is connected to a network TAP or is on a SPAN port and starts collecting ICS network traffic using passive (agentless) monitoring in no time. There is no impact on network deployment on OT networks because the sensor is not placed in the data path and only performs a passive scan of OT devices using proprietary **deep packet inspection (DPI)**.

Here are a few examples:

- An appliance (physical or virtual) can be placed in the shop floor **demilitarized zone (DMZ)** layer to bring visibility for monitoring all the shop floor cell traffic on a single glass pane.
- Another way is to place sensors in each shop floor segment. These sensors can be configured for local management or, with cloud management, can be placed in the shop floor DMZ section. A separate appliance (virtual or physical) can monitor the traffic in the shop floor DMZ section for SCADA, historians, or **manufacturing execution systems (MES)**.

In this section, we understood how the MDIoT sensor should be positioned after considering all the security requirements. It helps to place the sensor at the appropriate location in the network and configure monitoring ports based on the network architecture and security requirements.

OT sensor cloud connection methods

In the previous section, we learned about sensor placements. Now, let us move forward to understand how to monitor and maintain MDIoT. Connecting OT sensors to the cloud, and thus to MDIoT, is an important step to consider.

All the methods we are about to mention here emphasize the following:

- **Easy deployment**: Nothing extra to add to the Azure portal, especially to get the connection between the sensor and MDIoT going

- **Enhanced security**: No extra resource security needs to be turned on in **Azure Virtual Network (Azure VNet)**

- **Improved scalability**: As this is a SaaS-based solution, scalability can easily be achieved

- **Flexible connectivity**: The connectivity options will be discussed shortly, and you will see that the variety of options provided offers flexibility

Azure proxy

Azure VNet provides you with a proxy, which can aid in the process of connecting OT sensors to the MDIoT portal. The confidentiality of all the communications that occur between the sensor and the MDIoT portal (Azure) is ensured by the proxy.

In *Figure 5.11*, we can see that the OT environment sends data through the IT environment, as the IT environment is connected to the VPN gateway:

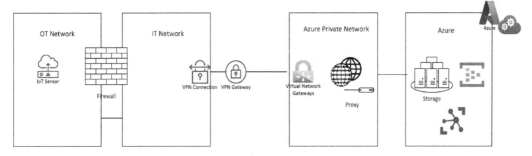

Figure 5.11 – A proxy connection to MDIoT with Azure VNet

The connection of the IT environment to the Azure cloud is via a VPN gateway in the IT network to virtual network gateways in Azure and a proxy in Azure for enhanced security. This connection, therefore, securely sends the OT data to the MDIoT portal for further analysis.

Proxy chaining

The layers in the Purdue model and the enterprise network hierarchy can help you connect to the MDIoT portal using multiple proxies. This is depicted in *Figure 5.12*:

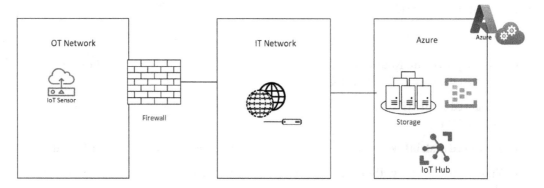

Figure 5.12 – A proxy connection to MDIoT with multiple proxies (proxy chaining)

This method allows you to connect the OT sensors to the portal, even without direct internet access. An SSL-encrypted tunnel is used to transfer data from the sensor to the portal using proxy servers.

The proxy server will not be maintained by MDIoT and needs to be installed and maintained by the organization itself.

Connecting directly

Another option is to directly connect the sensors to the MDIoT portal instead of traversing through the IT network.

In *Figure 5.13*, we can see that the sensor placed in the OT network establishes a connection directly (with the MDIoT portal) through the firewall:

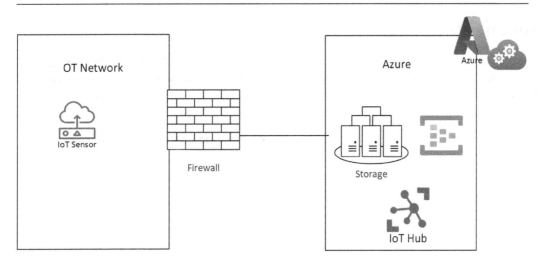

Figure 5.13 – Direct connection to MDIoT from the OT sensors

The connection to Azure (and then to the MDIoT portal) over the internet is secure, as it leverages **Transport Layer Security** (**TLS**) to encrypt the communication between Azure and the sensor. To ensure high security, it is always the sensors that initiate a connection with the Azure portal. Thus, this prohibits any inbound connections, saving you time and resources, as you do not need to configure firewall policies or inbound rules.

Multi-cloud connections

We live in a world where multi-cloud is a reality, so if you have to connect your sensors to MDIoT from other public clouds (AWS, Google, Oracle, etc.), this will be supported using one of the following methods:

- **Customer-managed routing through ExpressRoute**: Enterprises often leverage ExpressRoute for the many benefits it offers (speed, security, etc.) and the ability to use it to connect to the sensor from another public cloud. You can definitely look into using ExpressRoute (customer-managed) for routing traffic.

- **Cloud exchange provider's ExpressRoute**: Similar to the previous point, you can route traffic to the MDIoT portal from another public cloud through a cloud exchange provider's ExpressRoute.

- **A site-to-site (S2S) VPN over the internet**: Some organizations use an S2S VPN over ExpressRoute to route the network traffic, and the same can be used to route traffic toward the MDIoT portal as well.

So far, we have gotten a thorough idea of connecting the sensors in the OT environment to the MDIoT portal on Azure. Pick the one suitable for your organization or your customer's organization.

Summary

In this chapter, we have looked at the network topologies supported for MDIoT – that is, the ring and star topologies. While the ring topology creates a circular data path, the star topology is connected to a central hub to transfer the data. We also understood how the Purdue model connects to the MDIoT portal and learned the secure ways of connecting OT sensors to the MDIoT portal.

In the next chapter, we will see how the features of MDIoT help address open security challenges.

6

How Do the Microsoft Defender for IoT Features Help in Addressing Open Challenges?

We have seen in earlier chapters some of the challenges in securing organizations that follow **Industry 4.0** standards. These organizations reap the benefits of Industry 4.0 and the automation it brings forth, and forget to consider the security aspect.

In this chapter, we will focus on the areas that an IoT/OT organization might overlook when it comes to ensuring security and the impact this could have. At the end of this chapter, we will also learn about sensor installation, which is a starting step in your practical journey to MDIoT.

We will cover the following topics in this chapter:

- Missing asset inventory for IoT/OT devices
- Risk and vulnerability management
- Continuous IoT/OT threat monitoring, incident response, and threat intelligence
- The installation of the MDIoT service

Missing asset inventory for IoT/OT devices

When the assets of organizations are not tracked regularly and the organization does not know whether they are on the factory floor or in the manufacturing unit, this hampers productivity and leads to not-so-desirable business outcomes.

Organizations should place great emphasis on identifying what is in their network and the management of assets, including the current inventory of their business and software assets. The key to an effective cybersecurity strategy is making IT/OT asset management a priority. Most organizations, IT/OT teams, and cybersecurity teams struggle to understand what is in their network (assets), what is present on the device (what type of data it has, access to and from the device, and device criticality), who owns it (owners and users), and where the assets are located geographically and logically.

The following are some examples to put into perspective the havoc a *missing asset inventory* can create:

- Inability to identify unauthorized and rogue devices

- Being unable to distinguish **crown jewels** (the most important device, resource, and user with significant business value) from the rest of the assets may lead to the wrong evaluation of your OT cyber risk

- Vulnerability closure priority and response priority in case of an adverse situation

- Inaccuracy in reporting the implementation of security controls, threat intelligence, and incident response

You cannot protect what you do not know you have. Not knowing what is connected to your network may pose a risk. You may miss monitoring/responding to attacks on assets that belong to you but you don't know about.

A real-time asset inventory may help organizations to avoid incidents by being proactive in deploying security measures. A solid inventory of OT/IT/IoT assets becomes the foundation upon which to build a holistic cybersecurity program. Organizations may use this data as the basis of security. Take the following examples:

- In patch management, the inventory is the basis on which patches are identified. Vulnerability remediation can be prioritized if you know how important the asset is.

- Use threat intelligence to put proactive measures in place.

- Secure configurations are a must for security, but in order to maintain a secure configuration, organizations will require accurate, comprehensive, and current information about the inventory.

- Generate accurate reports for management regarding where you need to deploy a security control. A **what-if analysis** (a technique used to determine how expected performance is affected by changes in the assumptions on which the projections are based) helps in quantifying risk.

- Quickly identify unauthorized devices and take appropriate action.

A comprehensive, robust, and automated device/asset inventory improves cyber resilience. It's also the cornerstone of cybersecurity readiness for your OT ecosystem by enabling centralized inventory data to help identify, protect, detect, respond to, and recover in a **single-pane-of-glass** view (all details visible on a single portal). Organizations that have these capabilities feel more secure and in control. MDIoT allows you to see everything in your environment, with complete visibility of all IoT and OT assets and having rich context across each device, such as through communications, protocols, and behavior.

Risk and vulnerability management

Now that we know about the importance of an asset inventory, we need to talk about risks and vulnerabilities impacting OT/IoT assets. Common questions we hear from **Chief Information Security**

Officers (**CISOs**) and business teams are, *What are the risks for our crown jewels, that is, OT/IoT assets? What are the mitigation priorities for critical assets?*

In most OT/IoT environments, a vulnerability assessment is done on a fixed frequency (once a year or once every 6 months). This does not provide a real-time risk status for OT/IoT assets and overall business risk at any given point in time.

The MDIoT risk assessment report may help you here. The **risk assessment report** is a comprehensive vulnerability assessment report generated by MDIoT, based on network analytics using deep packet inspection, various behavioral modeling engines, and SCADA-specific state machine design. The good news is this is not a point-in-time truth; it is always current.

The following figure shows a sample screenshot of a risk report generated by MDIoT:

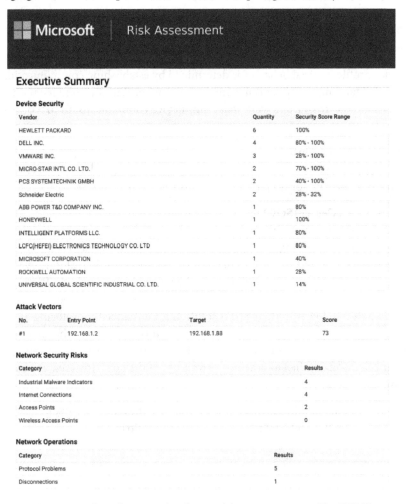

Figure 6.1 – Sample screenshot from a risk report generated by MDIoT

The report generates a security rating for each network device, as well as an overall network security rating. This rating is based on a calculation that aggregates device security ratings, associated vulnerabilities, configuration issues, and other network risks. The report provides mitigation recommendations to improve your current security assessment.

Continuous IoT/OT threat monitoring, incident response, and threat intelligence

IoT/OT threat intelligence and behavioral analysis are used to keep an eye out for unusual or unauthorized behavior. By immediately identifying unwanted remote access and rogue or compromised devices, you may improve IoT/OT zero-trust security. You can view traffic history, look up real-time alerts, and scan for risks quickly. You can also identify current dangers such as zero-day malware and survival techniques that static signs of compromise fail to pick up on. For more analysis, look into full-fidelity **packet captures (PCAPs)**.

For a given device profile, normal behavior is determined by establishing communication flows as baselines and understanding the systems it communicates with. With this understanding, policies can be aligned with a zero-trust framework that limits device communications to required systems and nothing else.

Continuous monitoring also helps organizations to be resilient in responding to untoward situations fast. MDIoT enables the security operations team to keep a watch on OT assets and provide all needed information, including PCAPs, for deep forensics to investigate any suspected breaches.

The **Security Operations Center (SOC)** may also benefit from using threat modeling. **Attack vector reports** are pictorial presentations of the vulnerability chaining of exploitable critical devices. These vulnerability chains could be leveraged by an adversary to get access to important network devices. Possible attack vectors for the critical devices in question are calculated in real time by attack vector simulation, based on connectivity, vulnerabilities, and the possible traversed path. This is very helpful for the continuous and real-time monitoring of the protection status of crown jewels. Organizations may want to integrate MDIoT into their SIEM solution for enhanced monitoring and response. Organizations may start the journey with isolated OT monitoring and, once they mature, integrate it with IT monitoring. Mature organizations may start integrating IT/OT on a single pane of glass. You can leverage Microsoft Sentinel as it detects threats out of the box. However, you may integrate MDIoT with any SIEM solution.

So far, we have learned about how MDIoT can help fill in gaps in the industry. Now, let us jump straight into getting started with setting up a free trial and installing MDIoT.

The installation of the MDIoT service

This is what is required for you to get started with MDIoT:

- A Microsoft Azure subscription and account
- Owner or Contributor access to the Azure subscription

For the test lab setup, you could configure all the roles and services with the global admin account itself. But if this is being planned on a production tenant, you might want to consider leveraging a role such as security reader, security admin, security contributor, or subscription owner to establish a separation of duties.

It would be wise to also consider the availability (i.e., traffic from all European regions is routed through West Europe, and all other regions are routed through the East US regional data center) of all the regions that MDIoT and the Azure IoT hub are available in and whether your data residency covers the region as specified in the compliance policy of your organization or the residing country.

The following are things to keep in mind when you plan to deploy MDIoT:

- Consider the network switches that would support monitoring of the traffic from a SPAN port and hardware appliances for **network traffic analysis (NTA)** sensors.

> **Note**
> For on-premises systems and air-gapped environments, activation, SSL/TLS certificate management, and managing passwords require admin privileges.

- Recollect the sensor placement methods we discussed in the previous chapter, *Chapter 5, How Does Microsoft Defender for IoT Fit into Your OT/IoT Environment/Architecture?*, and leverage them in designing your architecture. You can increase the number of devices in intervals of 100. Monitored devices are **committed devices**.
- Physical or virtual deployments are supported. You can also look into purchasing preconfigured and certified MDIoT appliances.
- Identify the switches, IPs, DNS servers, subnets, default gateways, hostnames, and so on. Configure port mirroring and connect the output port to the OT sensor.

Now, it's time to move on to onboarding the Azure subscription to collect the traffic and analyze the monitored data. Let us enable the trial plan (as an example here) for MDIoT:

1. Go to `portal.azure.com`.
2. Open the **Defender for IoT** page, as shown in *Figure 6.2*:

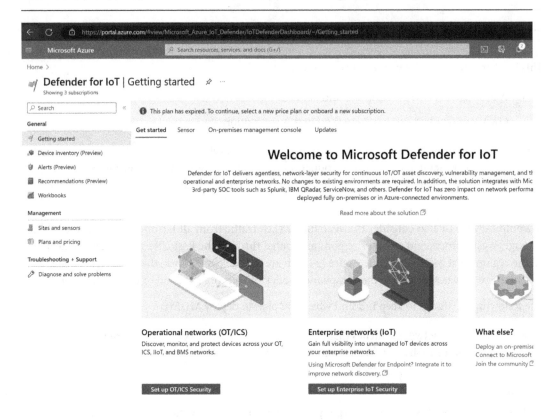

Figure 6.2 – Defender for IoT portal

3.   Select the **Plans and pricing** option situated on the left under the **Management** header.

4.   Click **Add plan**, as shown in *Figure 6.3*, after which a **Purchase** window will open:

Figure 6.3 – Screenshot of the MDIoT Pricing page in the Azure portal

You can choose to enable a trial of MDIoT, or you can choose the purchase method as a monthly or annual commitment. You may use it for up to 1,000 monitored devices for trial purposes for 30 days.

You need at least one sensor device to get started, and you must ensure that the device is connected to a SPAN port on a switch.

It is highly recommended to plan the required capacity for hardware resources available on the sensor **virtual machine (VM)**. *Table 6.1* provides hardware configuration recommendations and supported bandwidth for building a sensor VM:

Deployment	Corporate	Enterprise	SMB
Max Bandwidth	2.5 Gb/sec	800 Mb/sec	160 Mb/sec
Max Devices	12,000	10,000	800

Table 6.1 – Capacity planning for MDIoT sensor installation

Once this plan has been devised, let us move on to downloading the sensors. This is a required step for installation on VMs/physical devices and is to be done by you as opposed to purchasing a preconfigured appliance:

1. In the Azure portal, go to the **Defender for IoT** page and, after selecting **Getting started** on the left, select **Sensor**, as shown in *Figure 6.4*:

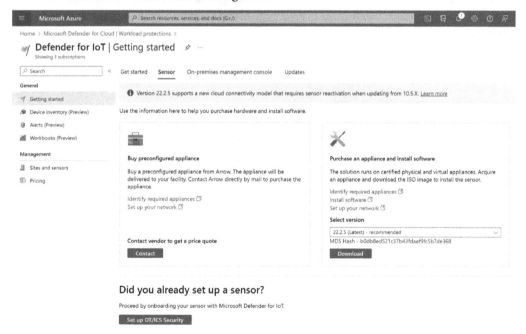

Figure 6.4 – Screenshot of the Sensor page in the MDIoT portal

2. You will now see two options:

- **Buy preconfigured appliance**: If this is what you would like to do, and you want to avoid manual configuration, please proceed.

- **Purchase an appliance and install software**: If you do identify any physical or virtual appliances in your OT environment, then go ahead with installing the sensor with the latest and recommended version (22.2.5 is the latest and recommended version at the time of writing this book). The downloaded sensor should be accessible from the device you plan to install it on.

> **Note**
> All downloaded media from the Azure portal is signed by the root of trust and the devices use signed/trusted assets only.

Let us now move on to creating a VM for the sensor software. This step will depend on whether you are using VMware or Hyper-V for installing the sensor VM.

Start by naming your VM and selecting the ISO you downloaded. Ensure you select the hardware profile of the VM based on what you deemed most suitable during the capacity planning stage.

Next, we will go ahead with turning on the VM and start installing the sensor.

The VM will start with the selected ISO image (sensor image), and you can proceed with selecting the language (in our example, English), as in *Figure 6.5*. MDIoT supports multiple languages, and you may select the one that suits your needs:

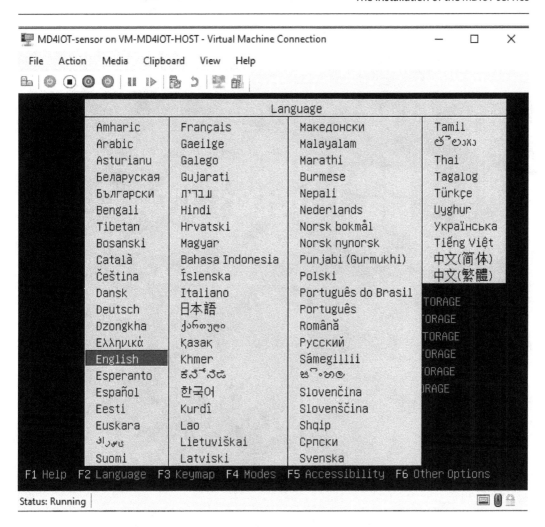

Figure 6.5 – Language selection prompt while installing the sensor

Based on your earlier capacity planning, select the size of the image (e.g., 4 CPUs, 8 GB RAM, and 100 GB storage are good enough for a proof of concept or lab), as highlighted in *Figure 6.6*:

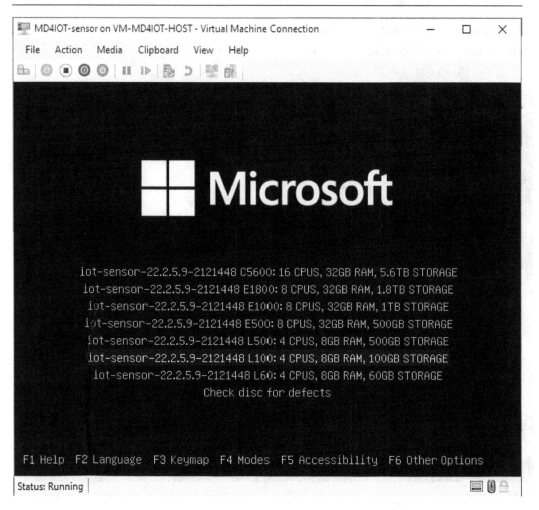

Figure 6.6 – Sensor installation – selecting the size of the image

In *Figure 6.7*, we can see the loading page for installing the image/sensor:

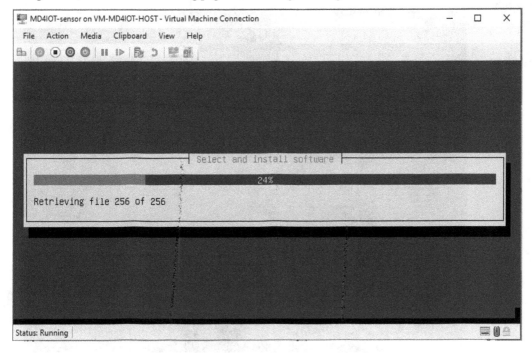

Figure 6.7 – Sensor installation – sensor loading

In *Figure 6.8*, you can see that we have leveraged Hyper-V to set up the sensor and that the sensor installation is in progress:

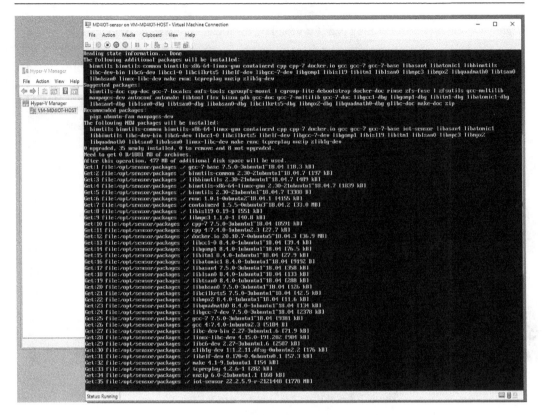

Figure 6.8 – Sensor installation – during the installation

In *Figure 6.9*, we can see that the grub file from the Linux server is created; remember that our sensor is purpose-built for Linux OS:

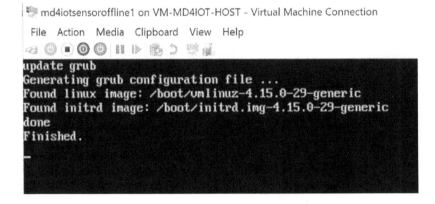

Figure 6.9 – Sensor installation – generating a grub config file

Network parameters, such as the management network IP address, subnet mask, appliance hostname, DNS, default gateway, and input interfaces, are selected based on the reference OT architecture the sensor is placed in.

Select eth0 (or eth1 based on your configuration) as your Ethernet connection by hitting the space bar, and then click **Ok**, as shown in *Figure 6.10*:

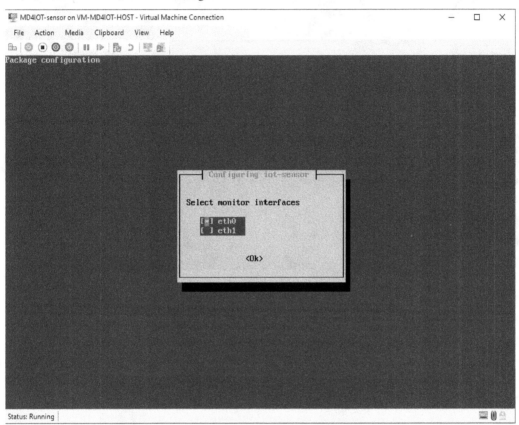

Figure 6.10 – Sensor installation – Ethernet port selection

If you do plan to connect your setup to erspan, you can select eth0 and click **Ok**. In the following prompt, you will be asked to select the IP address of your sensor. This is represented in *Figure 6.11*:

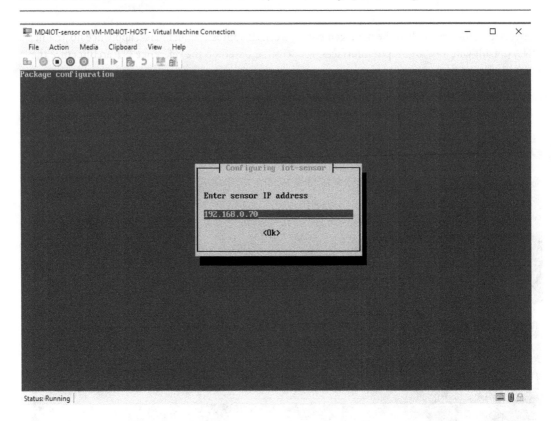

Figure 6.11 – Sensor installation – IP address of the sensor

Next, you will see a prompt with the location selected for backups. Click **Ok** on this prompt if you think the backup location looks good. In the following prompt, you can select the subnet mask, again based on your network requirements. We have chosen the subnet mask shown in *Figure 6.12* based on the lab requirements:

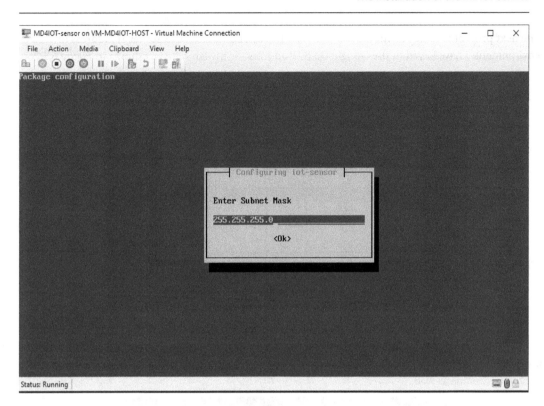

Figure 6.12 – Sensor installation – subnet mask of the sensor

The default gateway will be a deciding factor for whether you will make your sensor an online or offline sensor. For an offline sensor, key in any wrong value. *Figure 6.13* provides a screenshot of the gateway:

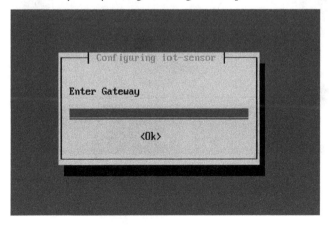

Figure 6.13 – Sensor installation – the gateway of the sensor

The next prompt is to key in the DNS server's IP. Please choose the DNS server IP address of your environment. I have keyed in the Google DNS IP address for my lab setup, as shown in *Figure 6.14*:

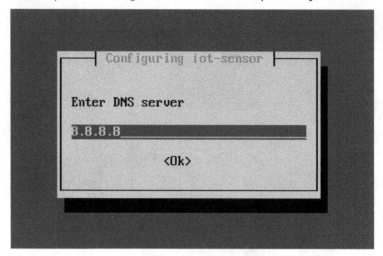

Figure 6.14 – Sensor installation – entering a DNS server IP

By default, the sensor hostname is as shown in *Figure 6.15*. However, you are free to change it to a customized name as well:

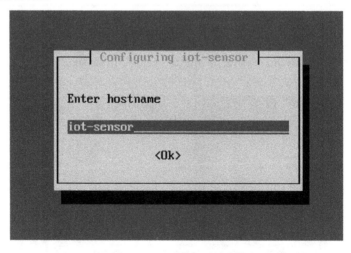

Figure 6.15 – Sensor installation – sensor name

For the lab setup, it was not important to set up any sensor as a proxy server. But in a real-world scenario, this is highly recommended. *Figure 6.16* depicts the option to convert the sensor into a proxy:

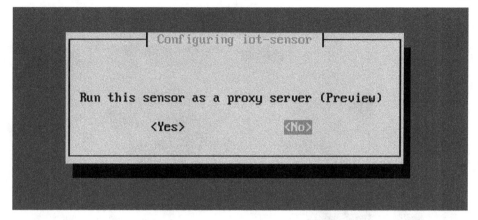

Figure 6.16 – Sensor installation – converting sensor into proxy server details

Always remember to take a screenshot of the password; if you skip this step, you will have to redo all the previously mentioned steps. *Figure 6.17* shows a screenshot of the username and password prompt:

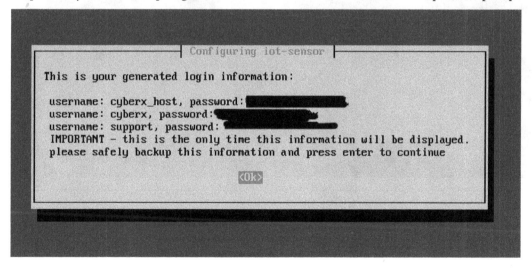

Figure 6.17 – Sensor installation – credentials page

Once we have all these details, the sensor installation goes through the final leg and finishes with the `install.sh` command, as shown in *Figure 6.18*:

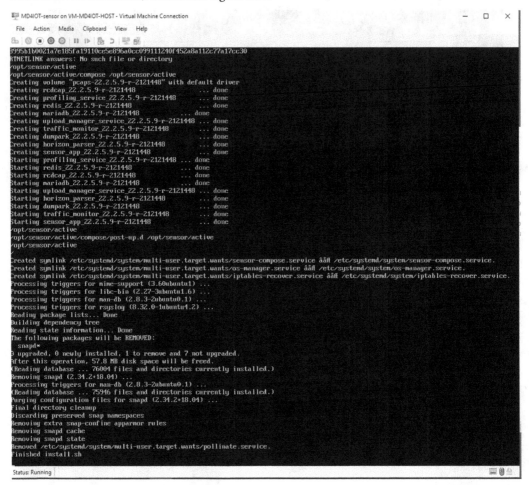

Figure 6.18 – Sensor installation – wrapping up the installation

Now, the VM will boot into Ubuntu OS and ask you to log in. Log in with the CyberX credentials that you saved earlier. This step can be seen in *Figure 6.19*:

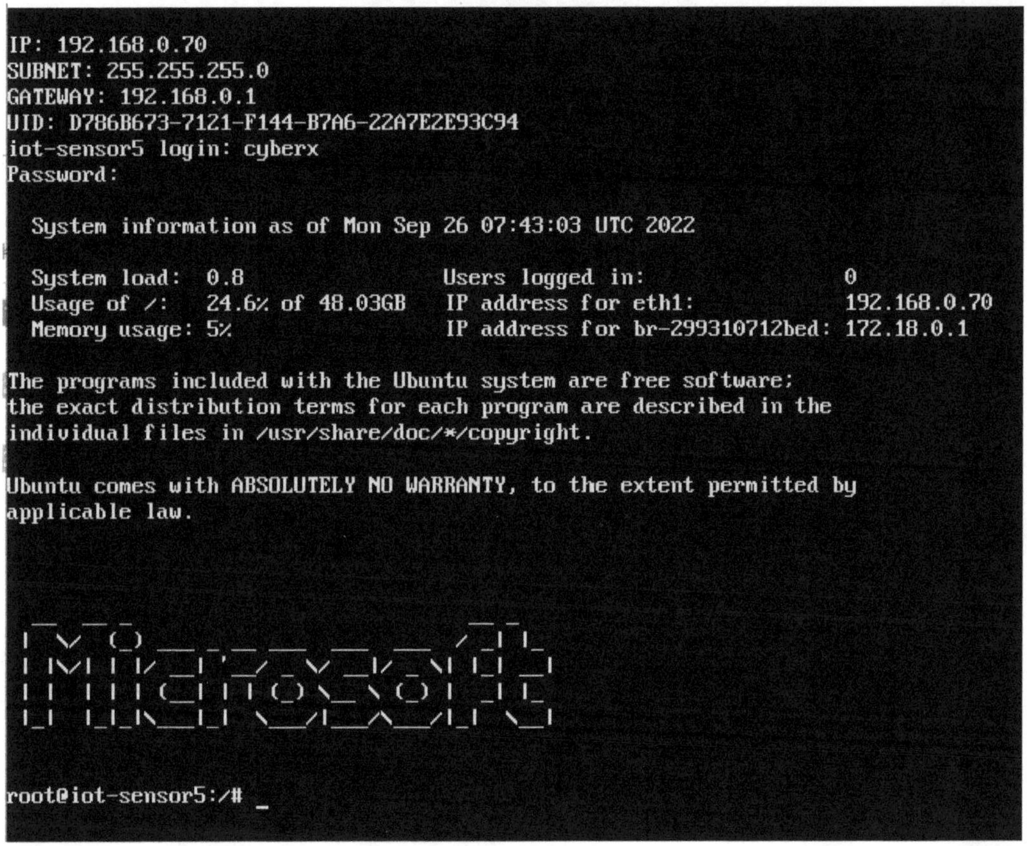

Figure 6.19 – Sensor installation – login console after the installation

And here we are: we've reached the end of the setup and are ready to log in to the sensor that we just installed. On the base server, go to a browser and key in the IP address of the sensor you chose or the IP address of the hardware appliance based on what you chose earlier. The sensor login page can be seen in *Figure 6.20*. Log in with the CyberX credentials that you saved earlier:

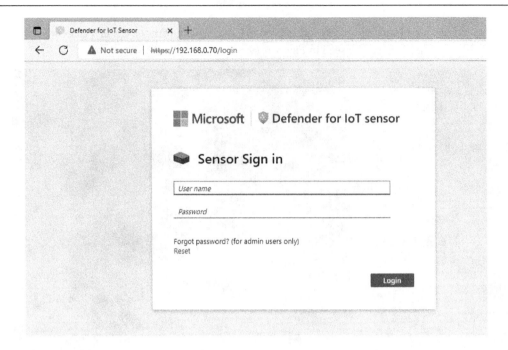

Figure 6.20 – Sensor login page

The next page will take you to **Sensor Network Settings**, as shown in *Figure 6.21*. Verify that the values you entered during the configuration match here and click **Next**:

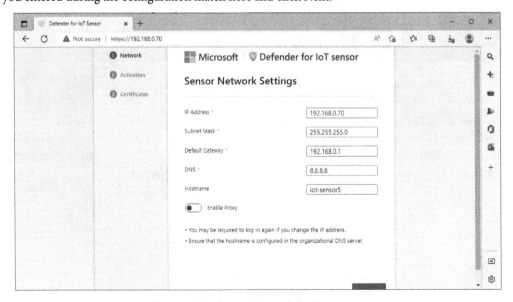

Figure 6.21 – Sensor Network Settings page

The next page requires that you upload the sensor activation file, as shown in *Figure 6.22*. How to get this activation file is explained in the next few screenshots. This file will activate your sensor based on whether you have enrolled in the free trial or a paid subscription:

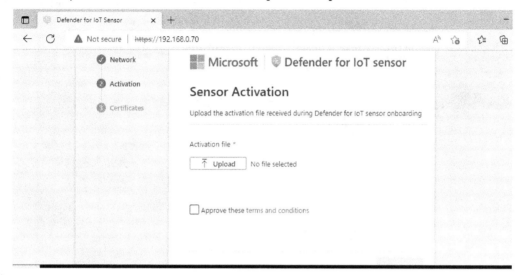

Figure 6.22 – Sensor Activation – getting the on-premises management console

To get the activation file and establish the communication for an online sensor, we need to register the sensor in the Azure portal.

Go to `portal.azure.com` and select **Defender for IoT**, then click on the **Sites and sensors** option, as shown in *Figure 6.23*:

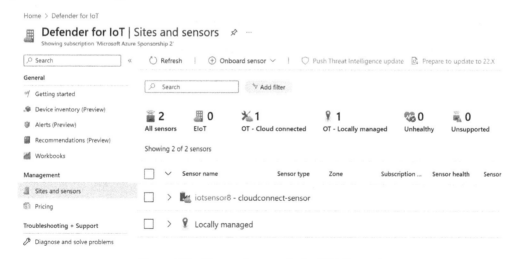

Figure 6.23 – Sites and sensors in the MDIoT portal

Since we have successfully completed *steps 1* and *2* – that is, configuring the sensor and the SPAN port or **test access point** (**TAP**), respectively – we can proceed with *step 3* to register the sensor.

Enter the sensor name and select the correct subscription for it. The **Cloud connected** toggle button can be turned on or off depending on whether you want an online or offline sensor.

Click **Register** to complete the registration, as shown in *Figure 6.24*:

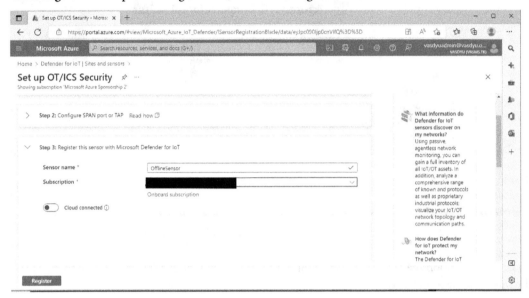

Figure 6.24 – Sensor registration in the Defender for IoT portal

On the next page, you will get a successfully registered sensor message, as shown in *Figure 6.25*.

At this point, you will also get the activation file that is required to activate the on-premises sensor.

Click **Finish** to end this registration:

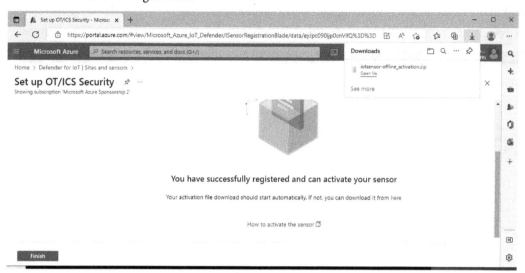

Figure 6.25 – Successful sensor registration

Heading back to our installed sensor now, do not forget to upload the activation package you just downloaded and proceed to the next step.

The next step (as depicted in *Figure 6.26*) will require you to enter the CA certificate details. This will ensure that the website is trusted each time you log in:

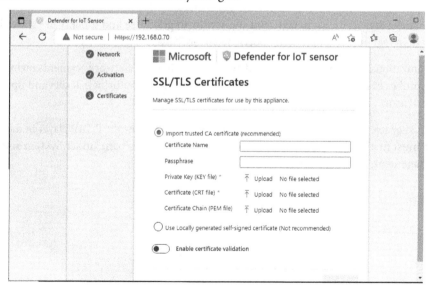

Figure 6.26 – Successful sensor registration

Finally, our first glimpse of the Defender for IoT on-premises portal is here, as in *Figure 6.27*. It is blank at the moment, as it is yet to capture any details/devices/alerts:

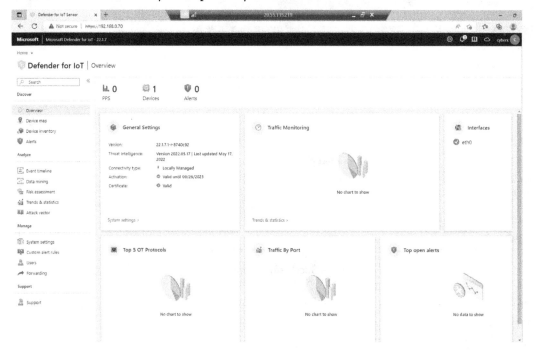

Figure 6.27 – Successful login to Defender for IoT on-premises portal

Now that we have logged in, we need to put the sensor to good use and quickly get started with capturing data in real time through SPAN ports, TAPs, or PCAPs.

In some organizations, it might be difficult to deploy sensors in some network segments owing to the size or the number of devices. In these situations, we recommend capturing packets and uploading them on the sensor.

We need to first enable the PCAPs setting in the sensor console. To show the PCAP player in the sensor console, we must first select **Pcaps** in the **Advanced configurations** pane under **System settings | Sensor management**, as shown in *Figure 6.28*:

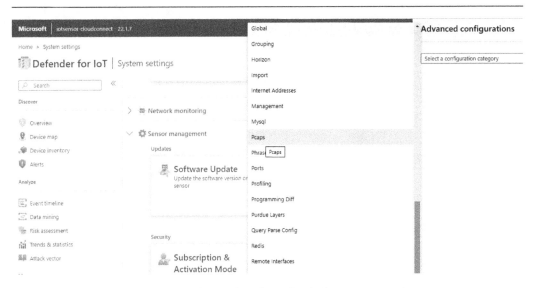

Figure 6.28 – Advanced configuration in the sensor console

Now, it is time to change the value of enabled to 1, as shown in *Figure 6.29*. This action is followed by clicking **Yes** under the warning and closing the prompt to save the changes:

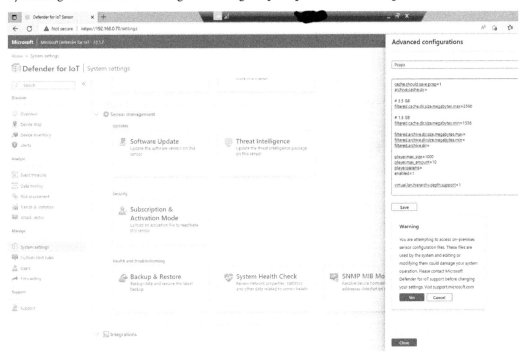

Figure 6.29 – Enabling PCAPs in the sensor console

Now, we are free to upload the PCAPs on the portal, as shown in *Figure 6.30*. You can bring the PCAPs and upload them on the sensor console:

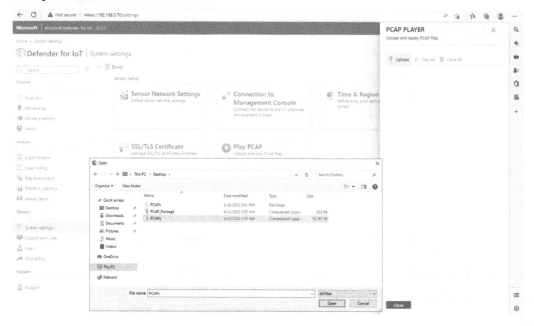

Figure 6.30 – Uploading PCAP files

Now, it is time to play all the PCAPs we have uploaded, as shown in *Figure 6.31*. This will ensure that MDIoT is able to parse through the traffic listing all the devices, alerting us of any abnormalities, and allowing us to assess risks:

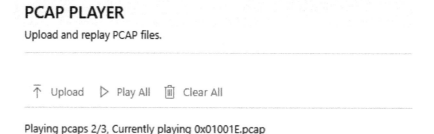

Figure 6.31 – Playing PCAPs on the sensor console

Here is a glimpse of the MDIoT portal (*Figure 6.32*) after we have played the PCAP files:

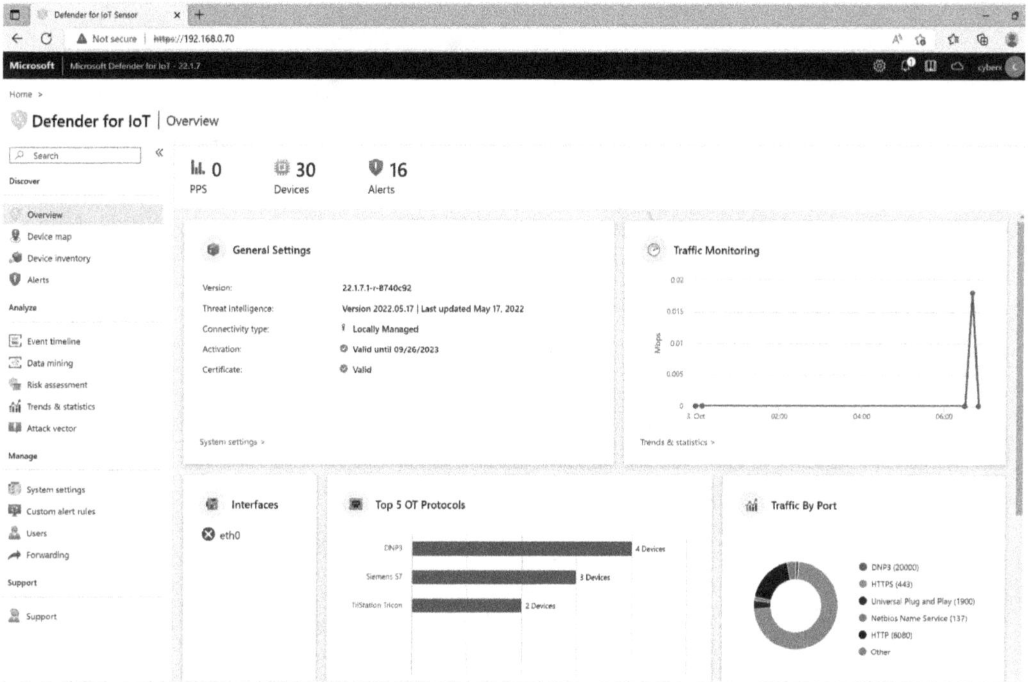

Figure 6.32 – Screenshot of the Defender for IoT portal after playing PCAP files

We will learn more about most of these components and features in the next few chapters. In this section, we have successfully gotten hands-on with setting up the MDIoT sensor and have even managed to parse some traffic to get an understanding of the data captured.

Summary

In this chapter, we have touched upon some of the important features of MDIoT and seen how it can help address open challenges by addressing the asset inventory, risk and vulnerability management, and continuous threat monitoring. We have also learned how to install the MDIoT sensor and seen it populating the dashboard with data.

It is now time for us to understand what the data we have collected so far can do for us as an organization and how these threat insights can save the day. We will look at this in the next chapter by understanding asset inventory. Asset inventory is a critical part of the OT/IoT cybersecurity landscape, and we will delve deeper into it next.

Part 3: Best Practices to Achieve Continuous Monitoring, Vulnerability Management, Threat Monitoring and Hunting, and to Align the Business Model Toward Zero Trust

In this section, the focus is on how continuous monitoring, asset inventory, vulnerability management, and threat monitoring help organizations implement the zero-trust model. We align this with the NIST security framework to achieve agility and better response by establishing best practices in the organization to monitor, secure, and manage the OT-IoT environment.

This section includes the following chapters:

- *Chapter 7, Asset Inventory*
- *Chapter 8, Continuous Monitoring*
- *Chapter 9, Vulnerability Management and Threat Monitoring*
- *Chapter 10, Zero Trust Architecture and the NIST Cybersecurity Framework*

7

Asset Inventory

We now have an understanding of how MDIoT aids in filling the cybersecurity gaps in OT/IoT organizations. In this chapter, we will understand how *identifying your assets is of paramount importance*. If you do not know the assets (IoT/OT) in your environment, you will not be able to protect them. Therefore, as we will see, increasing the visibility of your assets will help to reduce risk.

We will cover the following topics in this chapter:

- The device inventory in an on-premises console or the sensor console and the Azure portal
- Asset visibility – IoT/OT and identifying the crown jewels

The device inventory in an on-premises console or the sensor console and the Azure portal

An installed MDIoT sensor gathers devices into a on-prem portal as it scans through traffic. We call this the **device inventory**, and there are multiple places where this can be placed. There are three different places where you can examine the inventory, and which one to use depends on the maturity of the organization. An organization monitoring just one segment of the OT network can use the **sensor console** itself. An organization monitoring multiple isolated OT segments can use an **on-premises management console**, and the most advanced organization that wants to monitor IT, OT, IoT, and IIoT on a single console can use the Azure portal to see the consolidated inventory.

Often, organizations begin with the sensor console and move on to the MDIoT portal on Azure. However, cloud-native organizations start with Azure. In short, there is no rule of thumb, and from which console you decide to monitor the device inventory is based on where you are in your journey of protecting your OT devices.

Let us learn a little about each console:

- **The sensor console:** We can use the IP address assigned to the sensor console to log in to the sensor, as discussed in *Chapter 6*, *How Do the Microsoft Defender for IoT Features Help in Addressing Open Challenges?*.

- **The on-premises console**: The on-premises console consolidates all the IT and OT devices detected by sensors. These sensors further need to be connected to the on-premises console.

- **The Defender for IoT section in the Azure portal**: This brings together the device inventory from the cloud-connected sensors; OT, IT, and IoT devices are included.

This comprehensive view of all our assets and devices helps us to discover and identify them, aiding further in audits or troubleshooting.

Let us delve deeper into the device inventory from a sensor console perspective.

The sensor console

Log in to the sensor console that you have installed, using the procedure explained in *Chapter 6, How Do the Microsoft Defender for IoT Features Help in Addressing Open Challenges?*. Go to the **Device Inventory** tab, as shown in the following figure:

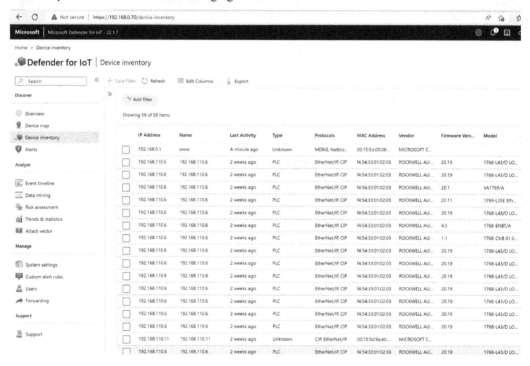

Figure 7.1 – Part of the device inventory from the sensor console

We can see that the device inventory is populated with all the devices the sensor has discovered. We can see details such as the device name, the IP address, the last known activity, the device type, protocols, the **Media Access Control (MAC)** address, the vendor, the installed firmware version, the model, and the operating system. You can also refresh the inventory to check whether a new device has been added.

You are free to add more columns according to your requirements. At least 15 more options are available on this list, as shown in *Figure 7.2*:

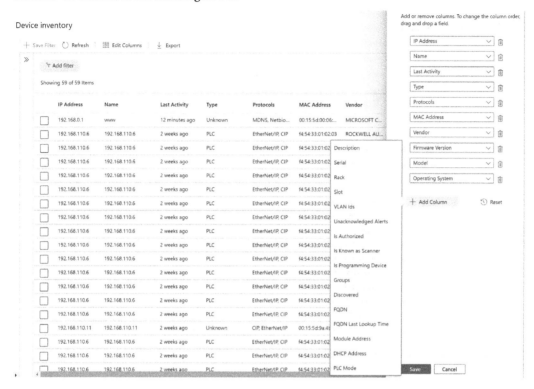

Figure 7.2 – Adding more columns to MDIoT's device inventory

If you want to export the device inventory in `.csv` format, you can click on the **Export** option and download the `.csv` file, as shown in *Figure 7.3*:

Figure 7.3 – Exporting the device inventory

The **Add filter** option in **Device inventory** lets you further filter down the devices shown based on your requirements. You can choose the column to filter using the options shown in *Figure 7.4*:

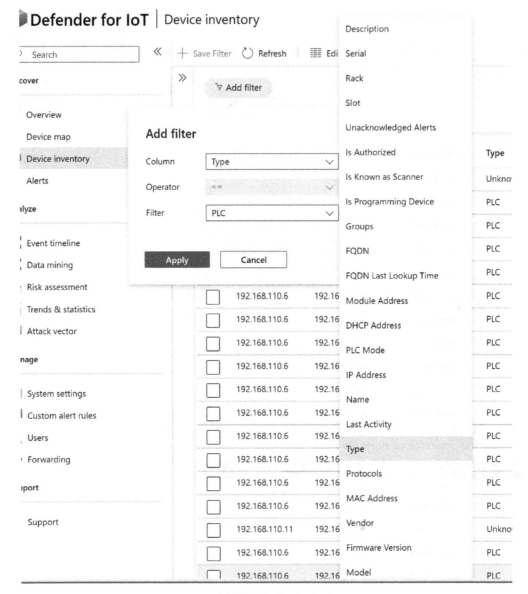

Figure 7.4 – Add filter in the device inventory

We can see that for **Column**, we have selected **Type**, and for **Filter**, we have selected **PLC**. This means that when the device inventory is applied, it will only populate devices that are of the *PLC type*.

In *Figure 7.5*, you can see that the filter is applied and you have the option to save it:

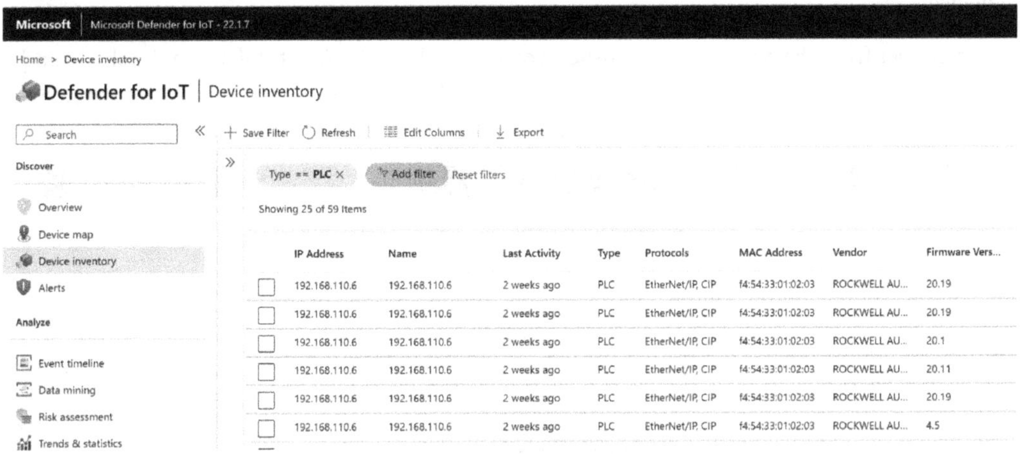

Figure 7.5 – Saving and resetting the filter

The advantage of saving a filter is being able to reuse it multiple times. After you have completed modifying the device inventory by applying a filter, you can also choose to reset the filter. Resetting it repopulates the inventory screen with all devices.

Let's highlight one of the devices in the device inventory, as shown in *Figure 7.6*:

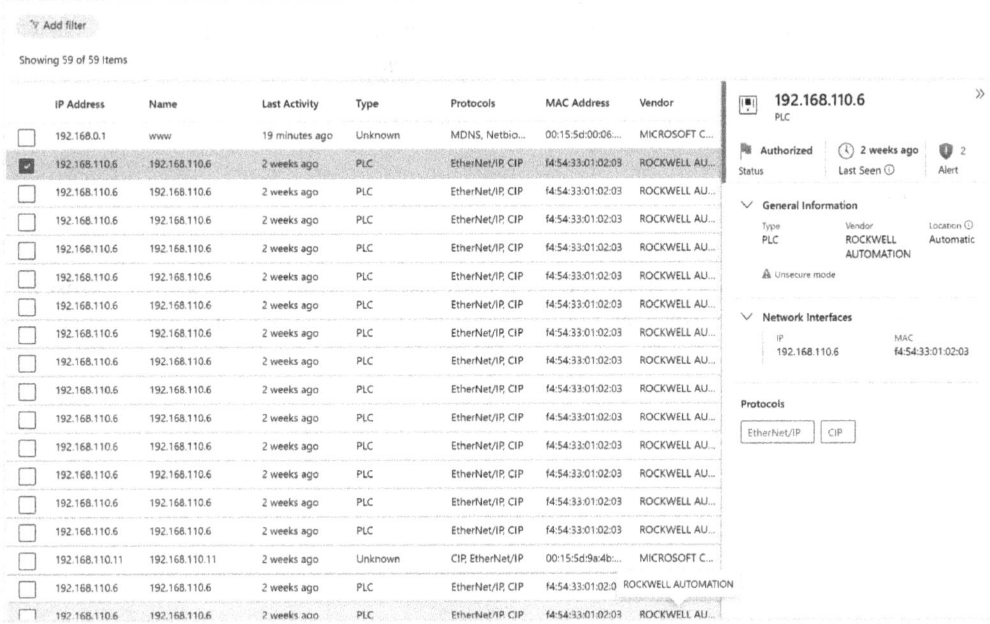

Figure 7.6 – Selecting a device in MDIoT's device inventory

Here, we can see details such as the type, the vendor, the authorized status, and the number of alerts on a device.

As you can infer from the preceding screenshot, you can click on the **Alerts** tab (with the shield and exclamation logo on the right of the preceding screenshot) to view the alerts present on the given device, as shown in *Figure 7.7*:

Figure 7.7 – Viewing the alerts on a device from the device inventory

As you might have already guessed, selecting the alerts further by clicking the alert name **EtherNet/IP CIP Service Request Failed** will provide you with more information on the chosen alert. We will discuss alerts in more detail in *Chapter 8*, *Continuous Monitoring*.

Now, let us dig into the device inventory in an on-premises console.

An on-premises console

In a real-world scenario, the number of sensors required to ensure the correct monitoring, discovery, and security of an OT environment is enormous. Organizations with multiple sites spread across the world need to have multi-sensor coverage, and a sensor every 80 meters (maximum distance between switches) is a requirement for complete coverage of all devices. To address this challenge and get a holistic view of all devices, it is recommended to connect all sensors to a single on-premises console (or as many on-premises consoles as possible, depending on your design requirements).

Let's take a quick look at the system settings of an offline sensor (the use and number of offline sensors will depend on the requirements of each organization):

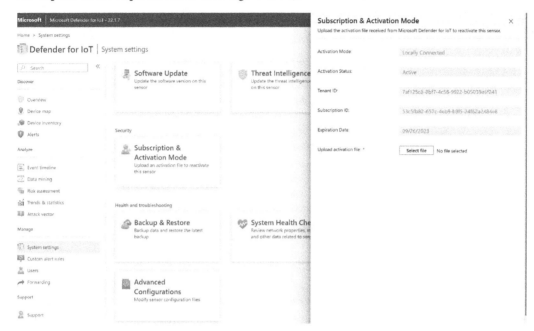

Figure 7.8 – Part of the locally connected sensor

As you can see in *Figure 7.8*, the **Subscription & Activation Mode** pane in **System settings** for a local sensor in the MDIoT portal shows that the **Activation Mode** setting is **Locally Connected**.

MDIoT in the Azure portal

As discussed previously in this chapter, whether an organization chooses MDIoT in the Azure portal depends on their location in the journey of protecting OT devices. For organizations choosing to leverage cloud-connected sensors, viewing and managing the device inventory on the MDIoT portal will be sufficient.

If you have not created a cloud-connected/online sensor and have yet to download the activation file required by your online sensor, *Figure 7.9* shows how to do so:

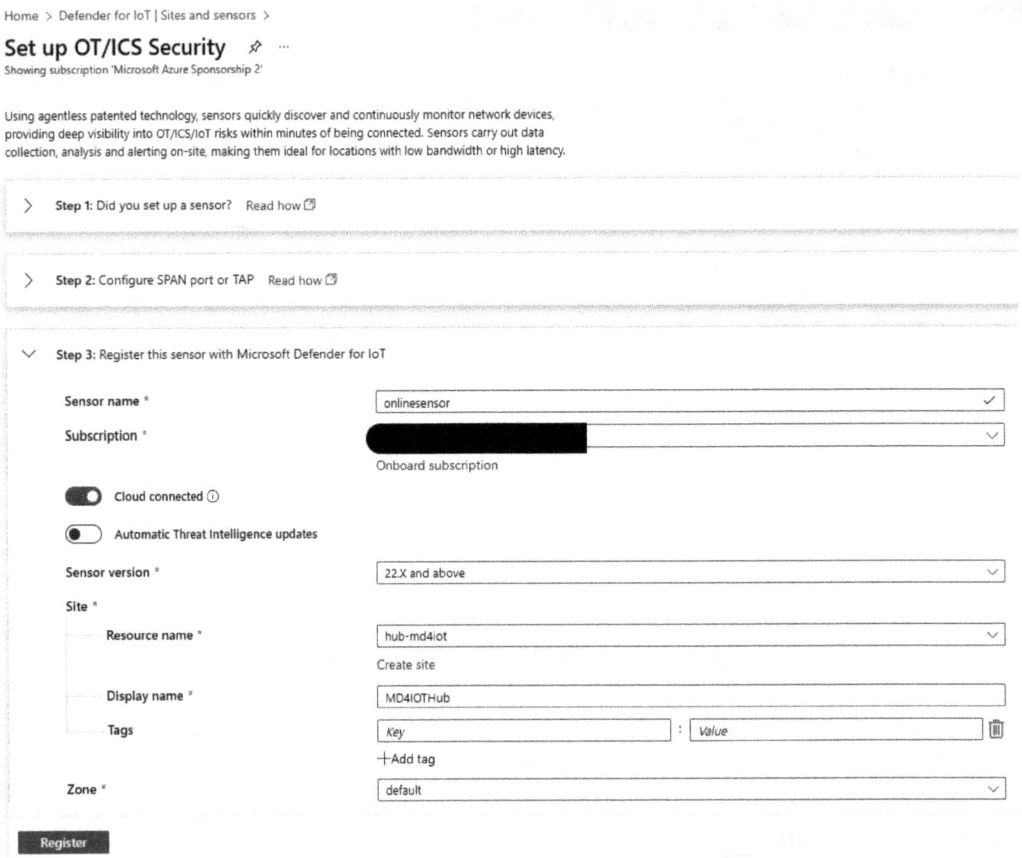

Home > Defender for IoT | Sites and sensors >

Set up OT/ICS Security 📌 ⋯
Showing subscription 'Microsoft Azure Sponsorship 2'

Using agentless patented technology, sensors quickly discover and continuously monitor network devices, providing deep visibility into OT/ICS/IoT risks within minutes of being connected. Sensors carry out data collection, analysis and alerting on-site, making them ideal for locations with low bandwidth or high latency.

> **Step 1:** Did you set up a sensor?   Read how 🗗

> **Step 2:** Configure SPAN port or TAP   Read how 🗗

∨ **Step 3:** Register this sensor with Microsoft Defender for IoT

Sensor name *	onlinesensor ✓
Subscription *	▮▮▮▮▮▮▮▮▮▮▮▮ ∨
	Onboard subscription

🔘 Cloud connected ⓘ

🔘 Automatic Threat Intelligence updates

Sensor version *	22.X and above ∨
Site *	
Resource name *	hub-md4iot ∨
	Create site
Display name *	MD4IOTHub
Tags	Key : Value 🗑
	+Add tag
Zone *	default ∨

Register

Figure 7.9 – Onboarding an online/cloud-connected sensor

With your sensor name keyed in and the **Cloud connected** option toggled on, you can register a new OT sensor. Upon successful registration, the activation file will be downloaded. These steps were previously covered when we discussed offline sensor activation in *Chapter 6, How Do the Microsoft Defender for IoT Features Help in Addressing Open Challenges?*.

You can choose to activate the sensor on the first login, or if you want to change the activation type from the sensor portal, you can do so from **System settings | Security | Subscription & Activation Mode**, as shown in *Figure 7.10*:

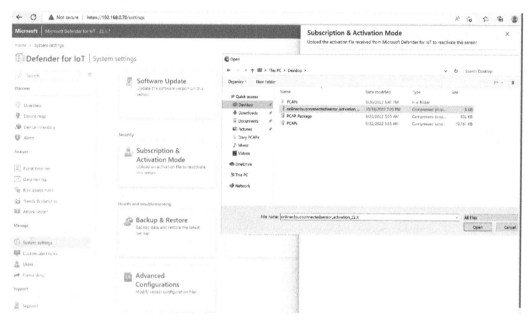

Figure 7.10 – Activation of a cloud-connected sensor from the sensor portal

The traffic generated is captured, analyzed, and sent over to the MDIoT portal in Azure. The **Device inventory** section of the MDIoT portal captures details about all the devices, as it did in the sensor portal. *Figure 7.11* represents all the devices that are captured by the cloud-connected sensors. It captures details such as the name of the device, the IP address, the type, the subtype, the last activity, the vendor, the model, and the MAC address. You can also filter the devices:

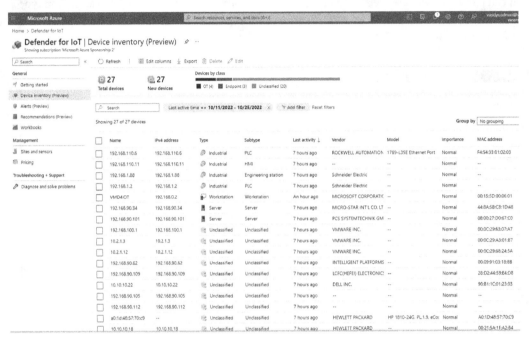

Figure 7.11 – Device inventory in the MDIoT section of the Azure portal

Filtering the device inventory works very similarly to what we did in the previous section for the sensor portal. *Figure 7.12* shows how to apply a filter by selecting the column type:

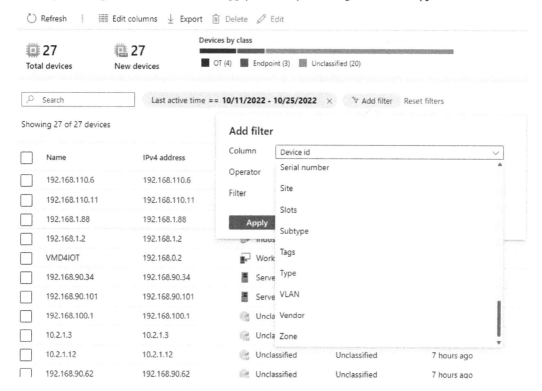

Figure 7.12 – Applying a filter to the device inventory

We can choose to group the device inventory based on protocols, Purdue level, site, vendor, sensor, and so on. In *Figure 7.13*, you can see the device inventory grouped by various protocols:

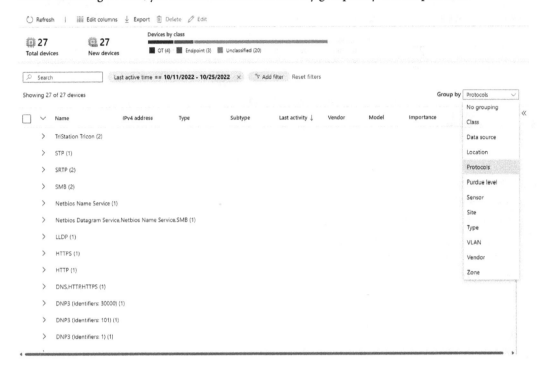

Figure 7.13 – Grouping of the device inventory

We can further select the device we are interested in and investigate it in more detail, as we can also see its alerts. *Figure 7.14* shows some brief information on the device such as its status, last seen details, and network interfaces:

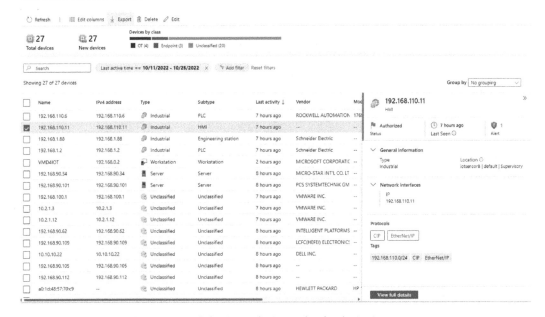

Figure 7.14 – Selecting a device under the device inventory

Clicking the **View full details** button at the bottom right of the preceding figure will take us through all the attribute information of the selected device. It also shows a glimpse of the device's vulnerabilities, alerts, and recommendations. We will cover these in the next few chapters. *Figure 7.15* shows detailed information gathered about the device:

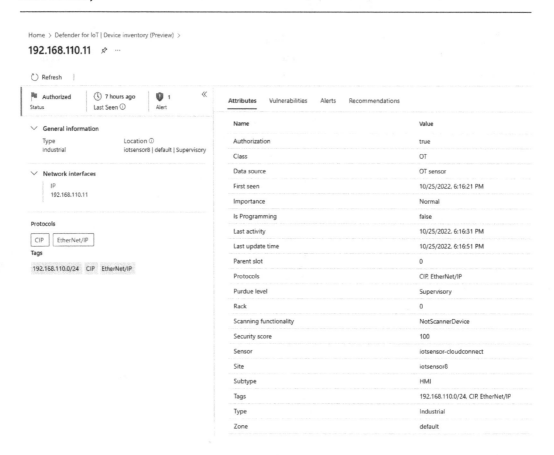

Figure 7.15 – Full details of the device in the device inventory

So far, we have learned about the device inventory in the sensor portal and the Azure portal. We have seen the benefits of tracking all the OT assets and being aware of attributes such as their vendors and last-active statuses. Also, associated IP addresses can give us additional information required to fully understand the assets.

Asset visibility – IoT/OT and identifying the crown jewels

One very important use of the device inventory is to identify the **crown jewels**. This aspect of asset inventory and classification helps if an adverse situation occurs, allowing you to define a strategy to prioritize the protection of your crown jewels. MDIoT allows you to mark important devices with ease. You simply go to the **Device map** tab, right-click on the device you want to prioritize, and click **Mark as important**, as shown in *Figure 7.16*. The devices that are marked with a star will then be considered important:

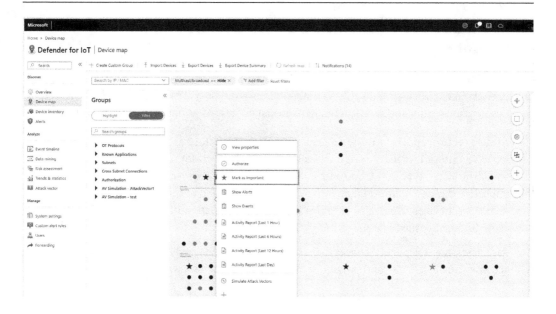

Figure 7.16 – Mark your devices as important

Organizations may want to enrich their asset inventories using other sources, such as **Configuration Management Databases (CMDBs)**, **Domain Name Servers (DNSs)**, firewalls, and web **Application Programming Interfaces (APIs)**, to enhance the data presented in the device inventory. An organization can benefit by using enhanced data to present information about the following things:

- Device purchase dates and warranty end dates

- Users in charge of a device

- Tickets opened for a device

- The last firmware upgrade date

- Whether internet access is allowed for a devices

- Whether an antivirus application is running on active devices

- Application details

- Users signed into devices

Important devices – generating attack vectors and risk assessment reports

When risk or attack vector reports are generated, the importance of devices is considered as one of the parameters in the calculation of risk. The suggestions given in risk assessment and attack vector reports also recommend using priority (i.e., devices that are tagged as important) to protect assets:

- **Attack vector reports**: All the devices marked as important are seen in the attack vector reports as **attack targets**. During attack simulations, attack vector reports produce information about the attack type and **Common Vulnerabilities and Exposures** (**CVE**), and they mark important devices as attack targets when they are connected to or surrounded by exposed or vulnerable devices. This helps organizations put appropriate mitigation controls in place to avoid future exposure.

- **Risk assessment reports**: Devices that are identified and marked as important are taken into account to calculate risk and provide security scores in risk assessment reports.

In this section, we have seen the importance of identifying the crown jewels of your assets and generating attack reports based on them. These generated reports provide you with a plethora of information about any potential vulnerabilities to attack in your environments, along with recommendations for counteracting them. We will talk in more detail about risk assessment and attack vector reports in *Chapter 9*.

Summary

In this chapter, we have learned the various ways to examine your organization's device inventory. The first way is by looking at information in the sensor portal. As a second option, you can forward this collected data to an on-premises portal for a collective view of all the data from multiple sensors. The third option is to connect the sensors to the cloud and send sensor data to the MDIoT portal on Azure. The information available about devices in the device inventory helps us manage and, of course, monitor assets. In the next chapter, we will learn about continuous device monitoring and how it helps those in the world of OT/IoT.

8
Continuous Monitoring

In the previous chapter, we started to understand the core features of MDIoT. Now, let us delve deeper into another of its features: continuous monitoring.

OT network sensors monitor the network traffic continuously across an organization's assets (for IoT and OT devices). This monitoring is done through the SPAN port or network TAP. This continuous monitoring is important in the field of cybersecurity, as it immediately picks up on any changes in the environment and reports anomalous behavior.

Continuous monitoring helps with the detection of policy violations, protocol violations, industrial malware, anomalies, and operational incidents. It helps in responding to the alerts generated by the system.

In this chapter, we will cover the following topics:

- The protocol violation detection engine
- The policy violation detection engine
- The industrial malware detection engine
- The anomaly detection engine
- The operational engine

The protocol violation detection engine

Generally, **protocol violations** can be identified by field values and packet structures that are being used in ways that go against the ICS protocol specifications.

In *Figure 8.1*, we can see **Modbus Exception** as an example of something picked up by the protocol violation engine. A secondary device did not send a response to the primary device when sending the exception code. This violation of the protocol was detected by MDIoT:

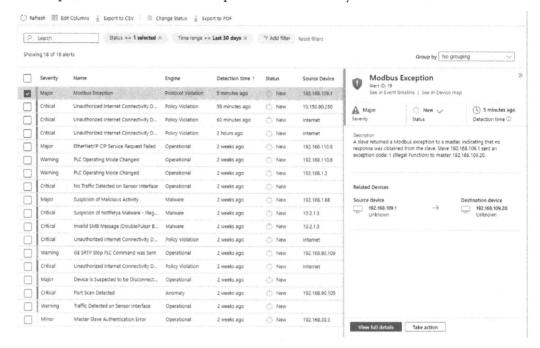

Figure 8.1 – Example of a protocol violation

> **Note**
>
> In the MDIoT portal for the Modbus alert, the primary device is referred to as the master device, and the secondary as the slave. However, due to their unsavory connotations dating back to colonization, standardization organizations have spoken against the usage of these terms. We have used primary and secondary in this book, but there are various alternatives available, such as controller-responder and primary-replica. You can read more about the history behind the original terms and replacements here: `https://cdm.link/2020/06/lets-dump-master-slave-terms/`.

In *Figure 8.2*, we can further see the protocol issues that were tracked and highlighted by MDIoT. Whichever protocols have deviated from a norm are detected and shown here:

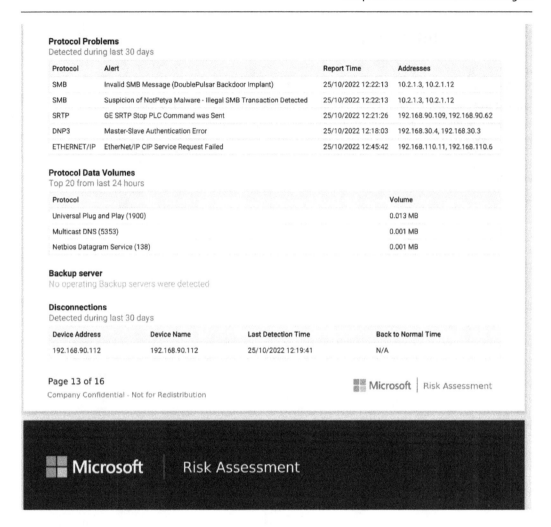

Protocol Problems
Detected during last 30 days

Protocol	Alert	Report Time	Addresses
SMB	Invalid SMB Message (DoublePulsar Backdoor Implant)	25/10/2022 12:22:13	10.2.1.3, 10.2.1.12
SMB	Suspicion of NotPetya Malware - Illegal SMB Transaction Detected	25/10/2022 12:22:13	10.2.1.3, 10.2.1.12
SRTP	GE SRTP Stop PLC Command was Sent	25/10/2022 12:21:26	192.168.90.109, 192.168.90.62
DNP3	Master-Slave Authentication Error	25/10/2022 12:18:03	192.168.30.4, 192.168.30.3
ETHERNET/IP	EtherNet/IP CIP Service Request Failed	25/10/2022 12:45:42	192.168.110.11, 192.168.110.6

Protocol Data Volumes
Top 20 from last 24 hours

Protocol	Volume
Universal Plug and Play (1900)	0.013 MB
Multicast DNS (5353)	0.001 MB
Netbios Datagram Service (138)	0.001 MB

Backup server
No operating Backup servers were detected

Disconnections
Detected during last 30 days

Device Address	Device Name	Last Detection Time	Back to Normal Time
192.168.90.112	192.168.90.112	25/10/2022 12:19:41	N/A

Page 13 of 16
Company Confidential - Not for Redistribution

Microsoft | Risk Assessment

Microsoft | Risk Assessment

Figure 8.2 – Example of protocol violation in a risk assessment report

Server Message Block (SMB), **Secure Real-Time Transfer Protocol (SRTP)**, **Distributed Network Protocol 3 (DNP3)**, and Ethernet/IP protocol changes are detected by MDIoT.

The policy violation detection engine

Any deviation from the learned baseline behavior or configured baseline behavior is detected by MDIoT and alerted on. In *Figure 8.3*, the **Unauthorized Internet Connectivity Detected** alert is seen, and from the name itself, we can understand that it is a breach of the baseline behavior, as the source is not authorized to communicate with the internet address:

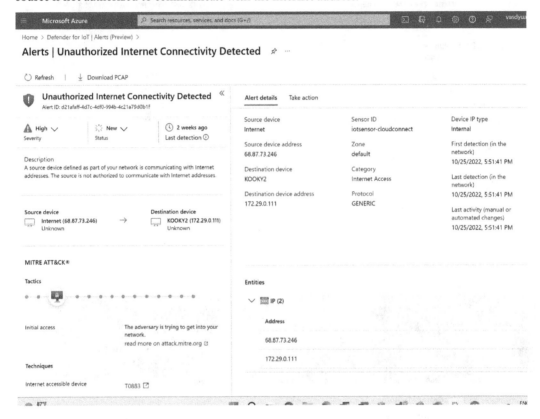

Figure 8.3 – Example of a policy violation

We will learn more about the structure of the alerts and their components in detail in *Chapter 9*.

The industrial malware detection engine

Understanding OT/IoT-based industrial malware is a must. The MDIoT detection engine contains this information to aid in better detection and alerting. Malicious activity on the network will be discovered by this engine.

In *Figure 8.4*, we can see the **Suspicion of Malicious Activity** pane in the MDIoT **Alerts** section. From the name itself, we understand that this attack could lead to exploitation by known malware – hence, further action is required from the **Security Operations Center (SOC)** or the admin team:

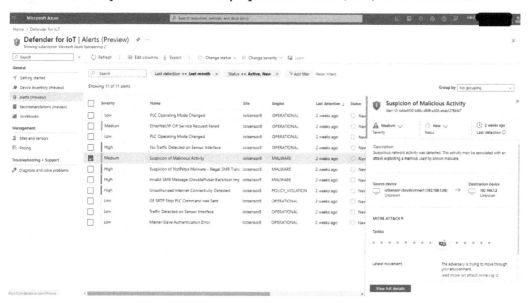

Figure 8.4 – The industrial malware detection engine

To gather further information about the malware, we can look into the full details of the **Suspicion of Malware Activity** alert:

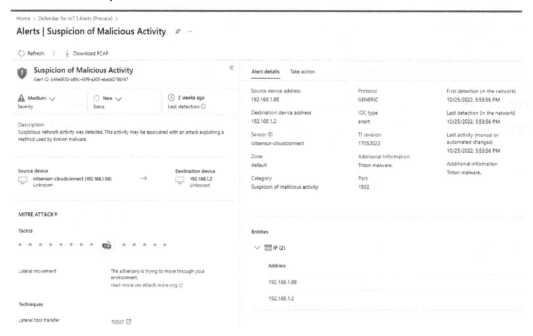

Figure 8.5 – Industrial malware detection engine: Triton malware

We can see that the name of the known malware is **Triton malware**. The lateral movement attempted by Triton malware is shown in the preceding figure.

The anomaly detection engine

Simply put, network behavioral anomalies are detected by the anomaly detection engine. In *Figure 8.6*, we can see an example of the anomaly detection engine at play. The **Port Scan Detected** pane depicts all the ports that were scanned by an attacker and the alert calls for immediate attention to the criticality of the incident. A **port scan** is a network anomaly and is detected by MDIoT:

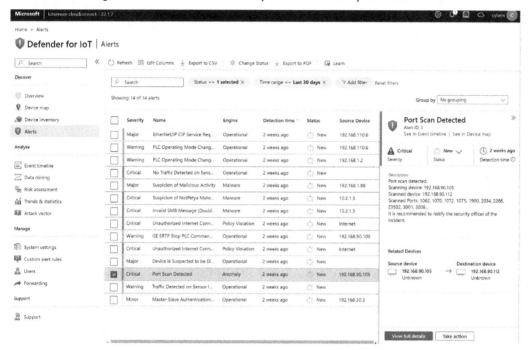

Figure 8.6 – Example of the anomaly detection engine at play

You can find out more about the alert and the anomaly by viewing the full details of the alert, as shown in *Figure 8.7*:

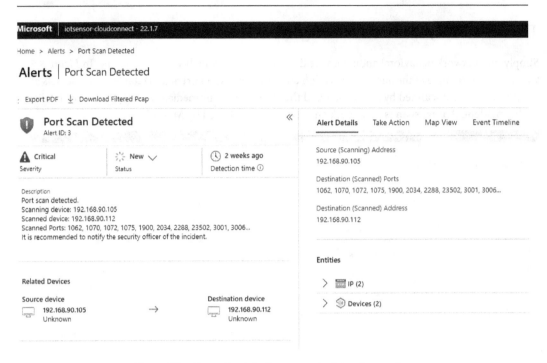

Figure 8.7 – Deep dive into the anomaly detection engine

You can find the details about the devices involved, including the **Media Access Control** (**MAC**) address, protocols, and vendor.

The operational engine

All malfunctioning entities and operational incidents are tracked by the operational engine.

In the example shown in *Figure 8.8*, we can see that the operating mode has changed, indicating that the PLC is not secure:

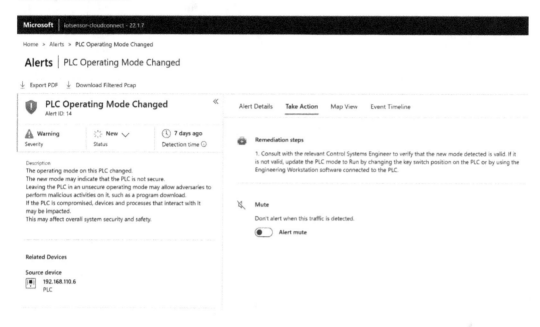

Figure 8.8 – Example of an operational violation

This change in the state might constitute a protocol violation as well.

Summary

Here, we got an understanding of the various engines that are part of MDIoT. All these detection engines contribute to the continuous monitoring of any threat vectors aimed at any organization's OT or IoT network or devices. These five engines provide ways to identify attacks and alert you about any attack coming the organization's way.

In the next chapter, we will interpret alerts in vulnerability management and threat monitoring.

Vulnerability Management and Threat Monitoring

In the previous chapters, we started learning about the core features of MDIoT. Let us continue in that spirit and move on to learning about risk and vulnerability management along with threat monitoring.

Risk management in IoT/OT incorporates identifying, analyzing, and controlling threats aimed at an organization's IoT/OT security. We learned about the challenges of cybersecurity in **Industry 4.0** in *Chapter 1*, and we have to ensure that the risks are known to the management and are assessed with risk mitigation steps in place.

Risk assessment forms an integral part of MDIoT. The most vulnerable devices, variations from the baselines, remediation priorities based on a secure score, network security risks, illegal traffic by firewall rules, connections to **Industrial Control System** (**ICS**) networks, internet connections, access points, industrial malware, indicators, unauthorized assets, weak firewall rules, network operations, protocol problems, backup servers, disconnections, IP networks, protocol data volumes, and attack vectors are some of the topics that will be discussed in this chapter.

In this chapter, we will discuss the following topic:

* Risk assessment

Risk assessment

Risk assessment provides you with a roadmap that will tell you what to protect and how to protect it. The ever-growing number of IoT or OT devices is increasing the security risks of organizations. Since an OT or IoT environment lacks modern controls such as auto-patching or a strong password (no default password), it becomes imperative that MDIoT should be leveraged, as it can instantly discover unknown vulnerabilities.

This understanding of vulnerabilities can be achieved by generating a risk assessment report. *Figure 9.1* shows the **Risk assessment** section in the MDIoT sensor portal. Click the **Generate report** tab:

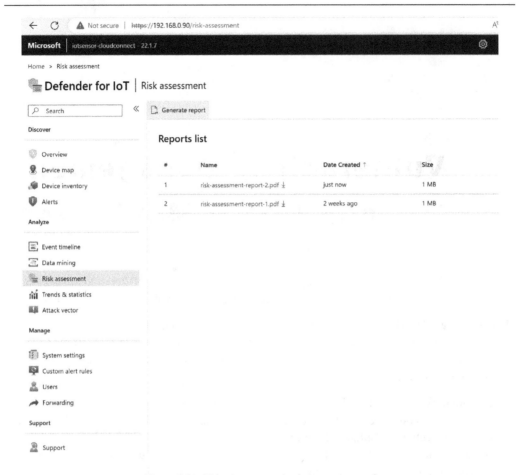

Figure 9.1 – Risk assessment in the sensor portal

When you click on a generated report, you will see the downloaded file, as shown in *Figure 9.2*:

Figure 9.2 – Downloading the risk assessment report

Let us move on to understanding what this report provides. *Figure 9.3* depicts the overall security status of your IoT or OT environment:

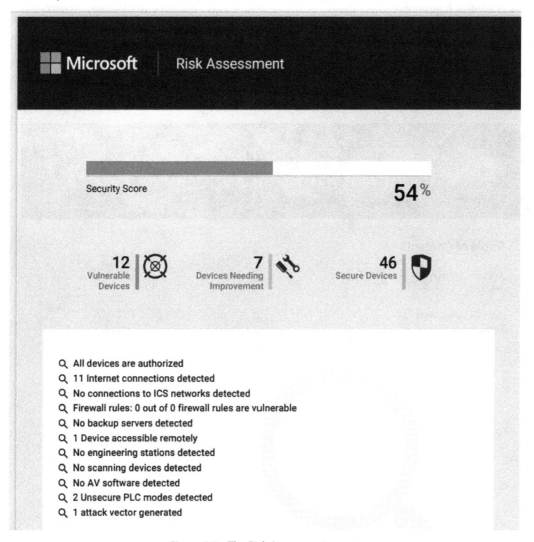

Figure 9.3 – The Risk Assessment report

The security score represents where your organization's security is at, and the entire report highlights the steps you can take toward improving this score. We can see that there are 12 vulnerable devices, 7 devices needing improvement, and 46 devices that are secure. This insight is crucial for any organization with OT or IoT devices to secure them further. And of course, the number of devices depicted in the report will vary with the actual report and the actual devices in the organization.

Figure 9.4 shows the table of contents for the entire report, and as learned previously (in *Chapter 4* as well as earlier in this chapter), we can see that the report is very exhaustive with information, such as **Top Vulnerable Devices**, **Network Security Risks**, **Network Operations**, **Attack Vector**, and **Mitigation**:

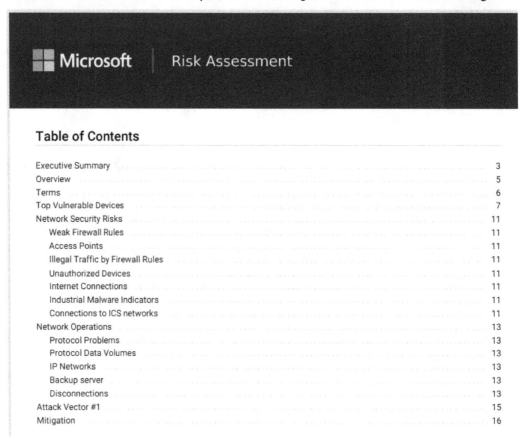

Table of Contents

Figure 9.4 – The Risk Assessment report – Table of Contents

The executive summary of the risk assessment report is shown in *Figure 9.5*:

Executive Summary

Device Security

Vendor	Quantity	Security Score Range
HEWLETT PACKARD	9	100%
DELL INC.	7	16% - 100%
YOKOGAWA DIGITAL COMPUTER CORPORATION	6	100%
FREEWAVE TECHNOLOGIES	5	100%
VMWARE INC.	4	28% - 100%
INTEL CORPORATE	2	100%
MICRO-STAR INT'L CO. LTD.	2	70% - 100%
PCS SYSTEMTECHNIK GMBH	2	40% - 100%
Schneider Electric	2	28% - 32%
ABB POWER T&D COMPANY INC.	1	80%
ARUBA A HEWLETT PACKARD ENTERPRISE COMPANY	1	100%
HONEYWELL	1	100%
INTELLIGENT PLATFORMS LLC.	1	80%
LCFC(HEFEI) ELECTRONICS TECHNOLOGY CO. LTD	1	80%
MICROSOFT CORPORATION	1	40%
PALO ALTO NETWORKS	1	40%
ROCKWELL AUTOMATION	1	28%
SIEMENS AG	1	100%
TELEMECANIQUE ELECTRIQUE	1	80%
UNIVERSAL GLOBAL SCIENTIFIC INDUSTRIAL CO. LTD.	1	14%
VIVAVIS AG	1	100%

Attack Vectors

No.	Entry Point	Target	Score
#1	192.168.1.2	192.168.1.88	73

Network Security Risks

Category	Results
Internet Connections	11
Access Points	5
Industrial Malware Indicators	4
Wireless Access Points	1

Figure 9.5 – Risk Assessment – Executive Summary

The depiction of **Device Security** is based on the device vendors you have in the environment. You will see the total quantity of each device per vendor and their respective security scores.

The **Attack Vectors** section throws light on the entry point and the target device in the organization and needs to be duly mitigated.

The **Network Security Risks** section provides information such as active internet connections, access points, industrial malware indicators, and wireless access points.

The risk assessment report details the steps you can take to mitigate the issues reported. *Figure 9.6* shows the **Mitigation** techniques:

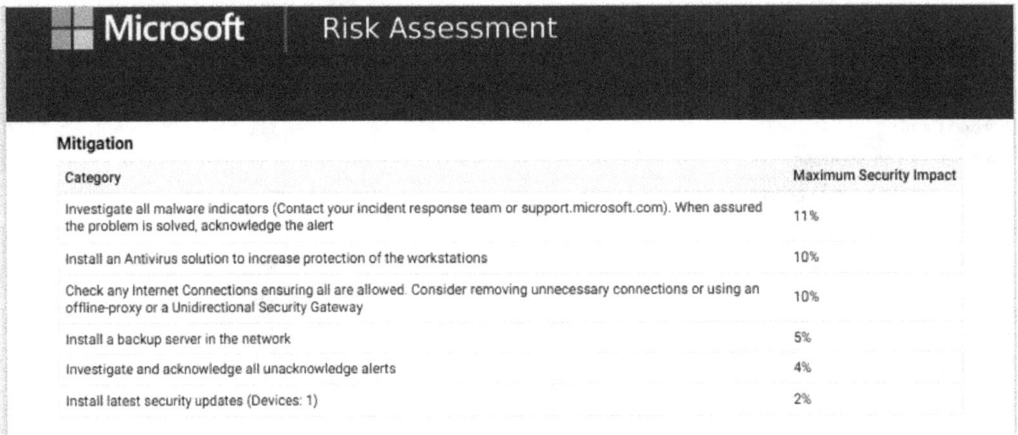

Microsoft | Risk Assessment

Mitigation

Category	Maximum Security Impact
Investigate all malware indicators (Contact your incident response team or support.microsoft.com). When assured the problem is solved, acknowledge the alert	11%
Install an Antivirus solution to increase protection of the workstations	10%
Check any Internet Connections ensuring all are allowed. Consider removing unnecessary connections or using an offline-proxy or a Unidirectional Security Gateway	10%
Install a backup server in the network	5%
Investigate and acknowledge all unacknowledge alerts	4%
Install latest security updates (Devices: 1)	2%

Figure 9.6 – Risk Assessment – Mitigation

You will also find the **Maximum Security Impact** percentage allocation for each mitigation technique. Upon mitigating each category, the mitigation technique with the highest percentage will improve your organization's security at a higher level. But it is always recommended that we focus on all the mitigations suggested for an overall security posture improvement in the OT/IoT world.

The next two pages of the 16-page risk assessment report further explain the inspection types and the modeling engine types (deep packet inspection, behavioral model engines, etc.). It uses this in the backend to provide us with a detailed report (this information remains the same for all reports). It discusses the security score for all network devices. It also contains the explanations or definitions of the terms used throughout the report, with clear information about when a device is marked as vulnerable (a security score < 70%) or when the device would be called secure (a security score > 90%).

As shown in *Figure 9.7*, this report gives you clear information about the most vulnerable devices in your organization's IoT or OT environment:

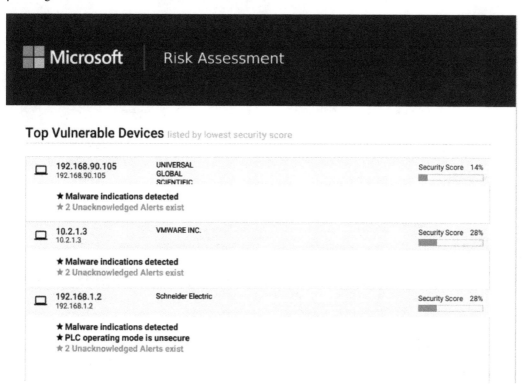

Figure 9.7 – Risk Assessment – Top Vulnerable Devices

This comprehensive list also indicates whether there is any malware and any unacknowledged, active alerts present. A security score of 14%–28% is way below the acceptable range. You should have a score of above 70% to not be marked as vulnerable. The report will also provide you with details about any ports in use, and also highlight the most severe **Common Vulnerabilities and Exposures (CVE)** that are unaddressed on the devices.

The **Network Security Risks** section in the risk assessment report gives information about weak firewall rules, illegal traffic according to firewall rules, unauthorized devices, and industrial malware indicators that were detected in the last 30 days, as shown in *Figure 9.8*:

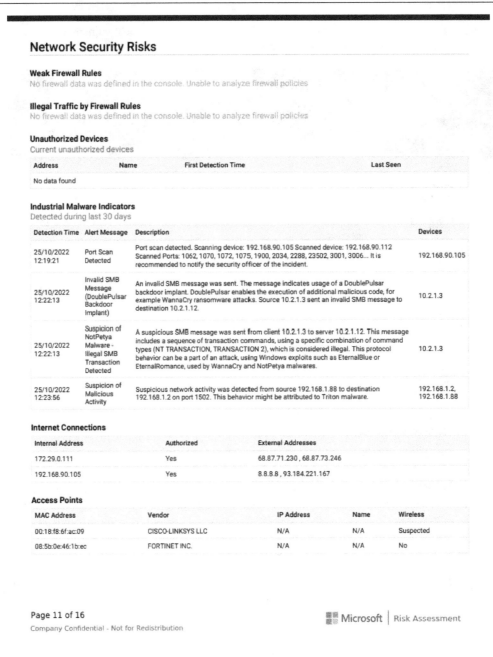

Network Security Risks

Weak Firewall Rules

No firewall data was defined in the console. Unable to analyze firewall policies

Illegal Traffic by Firewall Rules

No firewall data was defined in the console. Unable to analyze firewall policies

Unauthorized Devices

Current unauthorized devices

Address	Name	First Detection Time	Last Seen
No data found			

Industrial Malware Indicators

Detected during last 30 days

Detection Time	Alert Message	Description	Devices
25/10/2022 12:19:21	Port Scan Detected	Port scan detected. Scanning device: 192.168.90.105 Scanned device: 192.168.90.112 Scanned Ports: 1062, 1070, 1072, 1075, 1900, 2034, 2288, 23502, 3001, 3006... It is recommended to notify the security officer of the incident.	192.168.90.105
25/10/2022 12:22:13	Invalid SMB Message (DoublePulsar Backdoor Implant)	An invalid SMB message was sent. The message indicates usage of a DoublePulsar backdoor implant. DoublePulsar enables the execution of additional malicious code, for example WannaCry ransomware attacks. Source 10.2.1.3 sent an invalid SMB message to destination 10.2.1.12.	10.2.1.3
25/10/2022 12:22:13	Suspicion of NotPetya Malware - Illegal SMB Transaction Detected	A suspicious SMB message was sent from client 10.2.1.3 to server 10.2.1.12. This message includes a sequence of transaction commands, using a specific combination of command types (NT TRANSACTION, TRANSACTION 2), which is considered illegal. This protocol behavior can be a part of an attack, using Windows exploits such as EternalBlue or EternalRomance, used by WannaCry and NotPetya malwares.	10.2.1.3
25/10/2022 12:23:56	Suspicion of Malicious Activity	Suspicious network activity was detected from source 192.168.1.88 to destination 192.168.1.2 on port 1502. This behavior might be attributed to Triton malware.	192.168.1.2, 192.168.1.88

Internet Connections

Internal Address	Authorized	External Addresses
172.29.0.111	Yes	68.87.71.230 , 68.87.73.246
192.168.90.105	Yes	8.8.8.8 , 93.184.221.167

Access Points

MAC Address	Vendor	IP Address	Name	Wireless
00:18:f8:6f:ac:09	CISCO-LINKSYS LLC	N/A	N/A	Suspected
08:5b:0e:46:1b:ec	FORTINET INC.	N/A	N/A	No

Microsoft | Risk Assessment

Figure 9.8 – Risk Assessment – Network Security Risks

This basically saves you time from creating a comprehensive report that you can submit to the management team with all the extensive information. It also covers active internet connections from devices, access points, connections to ICS networks from devices, and so on.

Network Operations covers IP networks, protocol problems, protocol data volumes, backup servers (if any), and any disconnections, as shown in *Figure 9.9*:

Network Operations

IP Networks

Network	Mask	Name	Addresses
10.10.10.0	255.255.255.0	N/A	8
192.168.90.0	255.255.255.0	N/A	6
192.168.30.0	255.255.255.0	N/A	3
10.2.1.0	255.255.255.0	N/A	2
172.29.0.0	255.255.255.0	N/A	2
192.168.1.0	255.255.255.0	N/A	2
192.168.110.0	255.255.255.0	N/A	2
192.168.100.0	255.255.255.0	N/A	1
192.168.0.0	255.255.255.0	N/A	1

Protocol Problems
Detected during last 30 days

Protocol	Alert	Report Time	Addresses
SMB	Invalid SMB Message (DoublePulsar Backdoor Implant)	25/10/2022 12:22:13	10.2.1.3, 10.2.1.12
SMB	Suspicion of NotPetya Malware - Illegal SMB Transaction Detected	25/10/2022 12:22:13	10.2.1.3, 10.2.1.12
SRTP	GE SRTP Stop PLC Command was Sent	25/10/2022 12:21:26	192.168.90.109, 192.168.90.62
DNP3	Master-Slave Authentication Error	25/10/2022 12:18:03	192.168.30.4, 192.168.30.3
ETHERNET/IP	EtherNet/IP CIP Service Request Failed	25/10/2022 12:45:42	192.168.110.11, 192.168.110.6

Protocol Data Volumes
Top 20 from last 24 hours

Protocol	Volume
Universal Plug and Play (1900)	0.013 MB
Multicast DNS (5353)	0.001 MB
Netbios Datagram Service (138)	0.001 MB

Backup server
No operating Backup servers were detected

Disconnections
Detected during last 30 days

Device Address	Device Name	Last Detection Time	Back to Normal Time
192.168.90.112	192.168.90.112	25/10/2022 12:19:41	N/A

Figure 9.9 – Risk Assessment – Network Operations

Note all the IP ranges covered by sensors in this report. This holistic view is good to have and is scanned automatically and presented in a tabular column, with the total number of addresses under each network. The **Protocol Problems** section requires your immediate attention and needs to be fixed based on the endpoints reported by the risk assessment report. To find out whether any device has tampered with data volumes and whether you have any activity on the devices that has unwarranted data, you can check out the **Protocol Data Volumes** section.

Representing a vulnerability chain of endpoint devices in a graphical way is done in the **Attack Vector** section of the risk assessment report, as shown in *Figure 9.10*:

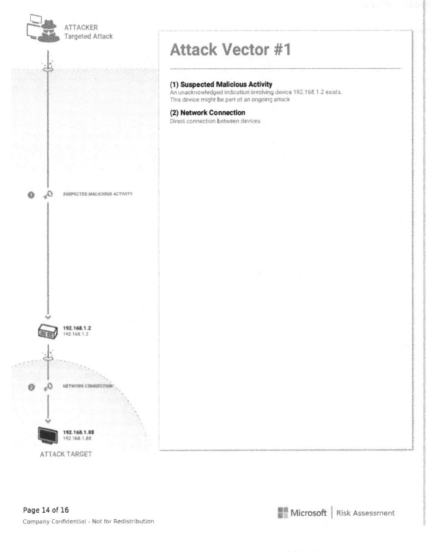

Figure 9.10 – Risk Assessment – Attack Vector

All attack vectors for a specific endpoint are calculated. Also, the mitigation activities conducted will improve the attack vector.

Moving on to **Mitigation**, we did see a little bit of it in the introduction of the report. Like the previous sections of the report, **Mitigation** provides detailed steps that we can take toward mitigating the threats, as shown in *Figure 9.11*:

Mitigation

Please note, the following enhancements are available:
★ Firewall policy import
★ Marking "important devices"
★ Further device information import

☐ Investigate all malware indicators (Contact your incident response team or support.microsoft.com). When assured the problem is solved, acknowledge the alert   **11%** Maximum Security Impact

Detection Time	Alert Message	Description	Devices
25/10/2022 12:19:21	Port Scan Detected	Port scan detected. Scanning device: 192.168.90.105 Scanned device: 192.168.90.112 Scanned Ports: 1062, 1070, 1072, 1075, 1900, 2034, 2288, 23502, 3001, 3006... It is recommended to notify the security officer of the incident.	192.168.90.105
25/10/2022 12:22:13	Invalid SMB Message (DoublePulsar Backdoor Implant)	An invalid SMB message was sent. The message indicates usage of a DoublePulsar backdoor implant. DoublePulsar enables the execution of additional malicious code, for example WannaCry ransomware attacks. Source 10.2.1.3 sent an invalid SMB message to destination 10.2.1.12.	10.2.1.3
25/10/2022 12:22:13	Suspicion of NotPetya Malware - Illegal SMB Transaction Detected	A suspicious SMB message was sent from client 10.2.1.3 to server 10.2.1.12. This message includes a sequence of transaction commands, using a specific combination of command types (NT TRANSACTION, TRANSACTION 2), which is considered illegal. This protocol behavior can be a part of an attack, using Windows exploits such as EternalBlue or EternalRomance, used by WannaCry and NotPetya malwares.	10.2.1.3
25/10/2022 12:23:56	Suspicion of Malicious Activity	Suspicious network activity was detected from source 192.168.1.88 to destination 192.168.1.2 on port 1502. This behavior might be attributed to Triton malware.	192.168.1.2, 192.168.1.88

☐ Install an Antivirus solution to increase protection of the workstations   **10%** Maximum Security Impact

☐ Check any Internet Connections ensuring all are allowed. Consider removing unnecessary connections or using an offline-proxy or a Unidirectional Security Gateway   **10%** Maximum Security Impact

Internal Address	Authorized	External Addresses
172.29.0.111	Yes	68.87.71.230 , 68.87.73.246
192.168.90.105	Yes	8.8.8.8 , 93.184.221.167

☐ Install a backup server in the network   **5%** Maximum Security Impact

☐ Investigate and acknowledge all unacknowledge alerts   **4%** Maximum Security Impact

Figure 9.11 – Risk Assessment – Mitigation Steps

You can follow the mitigation steps to decrease the security impact. Do not forget to acknowledge the alert when the incident is remediated. Ensure that security steps such as patching, firewall, and antivirus are regularly updated. Moreover, assessing any devices connecting to the internet – and if it is permissible to do so, in the environment – can all help mitigate the alerts in the environment.

We have a few more enhancements with the latest version of the sensor, and the report can be seen on the MDIoT in the Azure portal. When we go into the **Workbooks** section in the MDIoT portal, we can see that we have multiple workbooks available. Each one has its own benefits, and they let us focus on vulnerabilities. In *Figure 9.12*, we can see that the fourth workbook populated is **Vulnerabilities**:

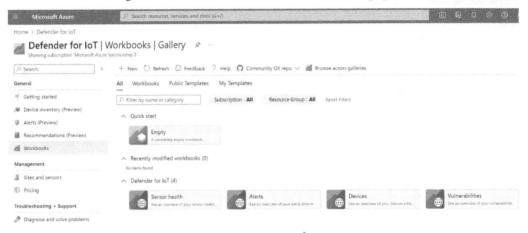

Figure 9.12 – Workbooks in the MDIoT portal

When we click on the **Vulnerabilities** workbook, it populates all the relevant dashboards such as **CVE by severity**, **CVE per site**, **CVE by vendor**, **Device vulnerabilities**, **Affected devices**, **Vulnerable devices**, and so on. Since these reports are populated on a cloud portal, it comes in handy for further analysis, and a quick glimpse can aid in a centralized investigation as well. Vulnerabilities across your OT environment are captured here, as shown in *Figure 9.13*:

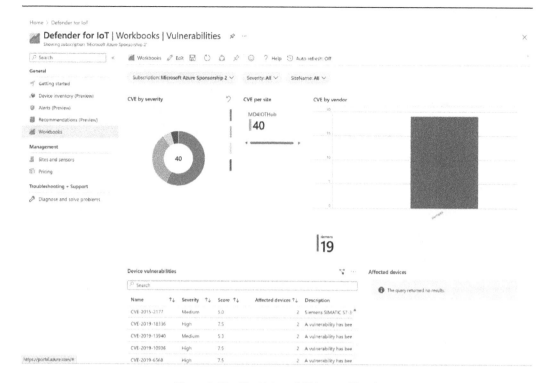

Figure 9.13 – The Vulnerabilities workbook

You must select the appropriate subscription to get the OT environment's highlights. And as can be inferred from *Figure 9.13*, the dashboards are editable.

When we click on the **Edit** button, we can choose the dashboard that we want to edit, which leads us to *Figure 9.14*:

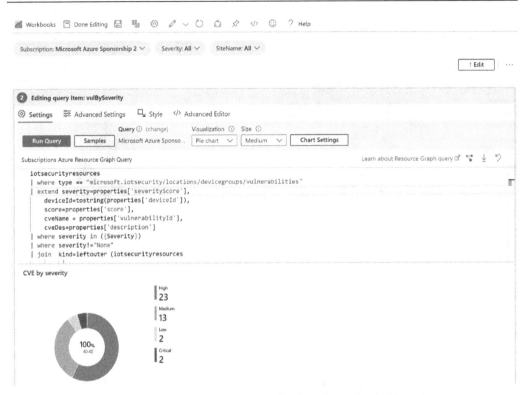

Figure 9.14 – The Vulnerabilities workbook – editing the dashboard

What we see here is a **Kusto** or **Azure Resource Graph** query. This is to indicate the power of the cloud and that of MDIoT to create reports based on a custom query and indicate vulnerabilities/threats better, tailored to your organization's needs.

Summary

In conclusion, we have learned about the various ways in which MDIoT aids in identifying threat/attack vectors and vulnerabilities. We have gotten a deeper understanding of the risk assessment report, which, in turn, shines a light on the security score of various devices and provides guidance toward mitigating them as well. To further increase usability and resourcefulness, we looked at the vulnerabilities highlighted in an Azure workbook as well, which aids in creating custom reports that can come in handy for your investigations. Use all of these to stay on top of your assets and devices, and secure them to thwart any attacks coming your way.

In the next chapter, we will explore **Enterprise IoT (EIoT)**, integrating MDIoT with **security information and event management** (**SIEM**) solutions such as Microsoft Sentinel (cloud-native SIEM), and the zero-trust approach regarding the IoT/OT industry.

10
Zero Trust Architecture and the NIST Cybersecurity Framework

By now, we know that **information technology (IT)** and **operational technology (OT)** are different, and the controls we apply to IT may not apply to OT. Though technological obstacles could make it difficult to adopt specific controls from **Zero Trust Architecture (ZTA)**, innovative thinking can assist businesses in applying ZTA concepts, even in delicate industrial settings.

In this chapter, we will cover the following topics:

- How MDIoT helps in implementing the NIST Cybersecurity Framework
- How MDIoT helps in ZTA implementation in an OT environment
- Validating ZTA with attack vectors

How MDIoT helps in implementing the NIST Cybersecurity Framework

The NIST **Cybersecurity Framework (CSF)** and ZTA are both frameworks aimed at improving cybersecurity, but they approach the problem from different angles.

The NIST CSF provides a set of guidelines and best practices for organizations to manage and reduce their cybersecurity risks. It covers the entire process of cybersecurity risk management, from understanding and reducing risks to responding to and recovering from incidents.

ZTA, on the other hand, is a security approach that assumes that no user, device, or service inside or outside an organization's network can be trusted by default and must be verified and authenticated before access is granted to resources. ZTA focuses on the trust aspect of security, while the NIST Framework provides a comprehensive approach to cybersecurity risk management.

The NIST CSF provides guidance for organizations to effectively manage their cybersecurity risks. It is a globally recognized and essential resource used by all sectors to understand, reduce, and communicate these risks. Despite changing cybersecurity threats, the CSF continues to effectively support risk management programs and improve communication within organizations. The framework has been adopted globally, both voluntarily and through government policies, due to its versatility and ability to address risks across industries, technologies, and borders. *Figure 10.1* depicts how MDIoT supports the NIST CSF:

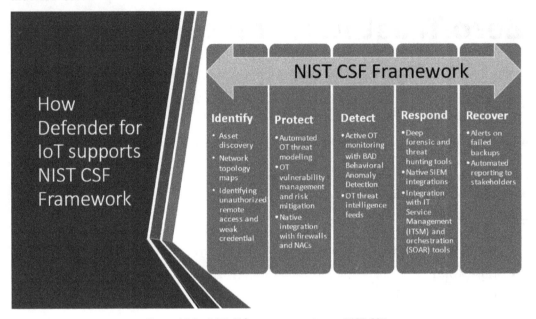

Figure 10.1 – MDIoT features mapping to NIST CSF

NIST's **National Cybersecurity Center of Excellence (NCCoE)** has published a 92-page report titled *Securing Manufacturing Industrial Control Systems: Behavioral Anomaly Detection* (NISTIR 8219).

The NIST report highlights the use of purpose-built IoT and **industrial control systems (ICS)** anomaly detection technologies (such as **CyberX**) as a means for manufacturing organizations to reduce the risk of disruptive cyberattacks, such as **WannaCry**, **NotPetya**, and **TRITON**. The report notes that these technologies allow organizations to effectively manage their cybersecurity risks without negatively impacting their OT networks. By using these anomaly detection technologies, organizations can detect and respond to cyber threats in real time, helping them to mitigate the impact of cyberattacks and protect their industrial processes.

The report provides guidance for organizations operating ICS in the manufacturing sector to enhance their cybersecurity posture. The report covers the implementation of the NIST CSF. It includes practical solutions, case studies, and implementation guides to help organizations understand, reduce, and manage their ICS risks and ensure the security of their industrial processes. The report is the result of collaboration between NIST, industry experts, and government agencies and highlights the importance of securing industrial control systems to protect critical infrastructure and prevent cyberattacks.

ZTA is a new security imperative. IoT devices, in particular, are highly attacked, paving the way to the internal network and the OT network as well. MDIoT, with a minimal **operating system** (**OS**) footprint, most commonly known as the **MDIoT micro agent**, can help organizations by providing rich information on risky OS configurations and a strong identity for authentication to determine anomalies and unauthorized activities.

How MDIoT helps in ZTA implementations in an OT environment

ZTA is not a one-size-fits-all framework. Instead, organizations need to create a tailored approach suitable for their business-specific needs that includes the people, processes, and technology needed to meet the objective. MDIoT is well-placed to help organizations to implement ZTA in the OT network. Starting with visibility, device health, vulnerability, and security, monitoring helps organizations start their journey toward ZTA confidently.

MDIoT is a non-intrusive and non-disruptive solution for existing networks, which is an essential requirement for critical OT systems and processes. This approach extends to ZTA services by monitoring network traffic and comparing observed behavior against specific baselines. It does not block unintended legitimate traffic. The organization can decide to flag identified ZTA access policy violations for further investigation or, with partner integration, can quarantine and block suspect devices and users if necessary. ZTA monitoring, which compares traffic patterns against established policies, is an important first step for most ZTA implementations to identify all network flows and all required application traffic critical to the application.

The ZTA principles of *never trust, always verify* connections, and assuming a bad actor is always present in the network (*assume breach*) help organizations be highly resilient and very agile against modern attacks. Likewise, MDIoT's focus on identifying assets, continuously verifying vulnerability assessments of endpoint devices, and understanding legitimate incumbent operations serves as an intelligent automated verification platform for every device in your organization, around the clock, seven days a week. We also have in-depth asset intelligence that understands the basic and expected behavior of OT devices or **programmable logic controllers** (**PLCs**). MDIoT can even identify when trusted devices may have been compromised and evaluated for quarantine or restricted access.

ZTA in an OT environment can be implemented in three steps:

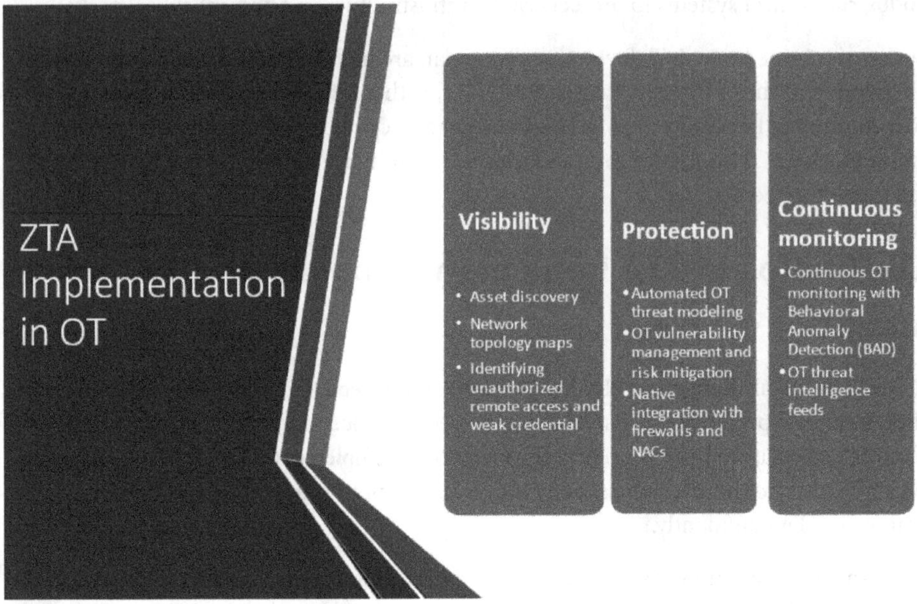

Figure 10.2 – ZTA implementation in OT

The three steps are as follows:

1. **Visibility**: Find and categorize assets that have a significant impact on operations, safety, and the business.

2. **Protection**: Utilize static and dynamic controls to isolate assets from unnecessary internet and production access.

3. **Continuous monitoring**: OT, IT, and IoT asset threat detection and response processes should be continuous, combined, and automated.

Visibility

MDIoT helps achieve complete visibility of all network devices. It leverages the asset inventory to create a baseline of authorized devices. It also allows administrators to set the importance of the devices.

Without knowing what types of assets you have, it is impossible to protect industrial assets or develop mitigating strategies. The security flaws of each asset type are different. Your network will be better protected from attacks if you have visibility of the assets you have, how they are set up, and how crucial they are to operations. MDIoT offers various methods to discover and bring visibility to your OT/IoT devices. With the given choices, you are empowered to select the one that suits your business requirements the best:

- **Agent-based**: This one is best for all the new devices built for your unique requirements.

  You can include security in new IoT gadgets and Azure IoT projects by using the MDIoT security agent. The MDIoT micro agent is compatible with common IoT OSs (such as Linux and Azure **RTOS** (which stands for **Real Time Operating System**)) and includes versatile deployment choices, such as the capacity to deploy as a binary package or edit the source code.

  The MDIoT micro agent offers unified security management by integrating with Microsoft's other security solutions and providing endpoint insight into security posture management and threat detection.

- **Network sensor-based**: OT network sensors employ agentless, patented technology to discover, educate, and continually monitor network devices for deep visibility into OT/ICS threats. Sensors are perfect for locations with little bandwidth or high latency because they gather, analyze, and alert users on-site.

- **Enterprise IoT**: Enterprise IoT enables visibility and security for IoT devices in a corporate ecosystem.

  Network protection provides coverage for all IoT devices in your environment by extending agentless functionality outside the operational segment. For instance, an enterprise IoT ecosystem might contain printers, cameras, and proprietary, custom-built devices.

Using any of the preceding methods or a combination of these methods helps you achieve the following benefits:

- **Save time and avoid human error**: Visibility brought about by the MDIoT inventory saves time by discovering all of your assets with minimal intervention. The best part is it eliminates possible human errors. With this, you get visibility of the assets and free up staff time to focus on other important tasks.

- **Audit trail**: Endpoint and other transient assets present a significant security risk to sensitive OT networks. You can audit which assets move around your system using real-time data provided by MDIoT. The equipment that connects to your network from vendors or other parties is also tracked. This information may be essential in pinpointing the point of an intrusion in the event of a cyberattack.

With this, we learned how MDIoT helps you to understand authorized devices and get visibility on newly added devices and the activity being performed by these devices to help you decide whether the device is safe or suspicious and then take action accordingly.

Protection

Mostly, the industrial network is one big implicit trust zone, depending on the existing OT network architecture. If possible, network segmentation can divide this trust zone into smaller, easier-to-manage chunks. However, the one question the OT team always ponders is: *Can you easily validate communications from different network segments?*

MDIoT helps you to address these questions confidently. As depicted in *Figure 10.3*, MDIoT helps you find communication between subnets, remote access, and internet connectivity from the devices identified:

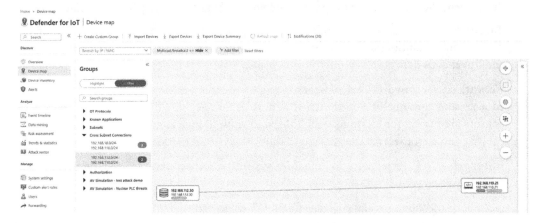

Figure 10.3 – Subnet 1 and subnet 2 ideally should not talk to each
other; however, it is found that the devices are connected

As depicted in the preceding screenshot, OT asset owners can safeguard themselves against improperly networked devices by learning about their assets and the networks they are connected to. Owners are now able to quickly address these issues thanks to improved real-time visibility.

The following screenshot depicts the internet connectivity in an OT network:

Figure 10.4 – Internet connectivity identified

This, too, may pose a threat to your OT network.

In the following screenshot, a remote connection is identified on the device:

Figure 10.5 – Remote connection

MDIoT allows you to identify any unwanted or malicious assets connected to your network, allowing you to take immediate action and begin remediation.

Continuous monitoring

When it's difficult to apply proactive controls closely, monitoring is key to resilience and responding quickly.

MDIoT helps you to monitor the alerts generated with real-time information about events that have been logged. MDIoT alerts help to improve network security and operations, such as the following:

- Unauthorized changes to device and network configurations
- Protocol and operational abnormalities
- Suspected malware behavior

Organizations may handle the alerts on the Azure portal to monitor, investigate, and act on alerts generated by MDIoT. Organizations may also integrate these with **security information and event management (SIEM)** solutions.

MDIoT natively connects to Microsoft Sentinel without much effort. However, the integration is not just limited to Microsoft Sentinel; it has the capability to integrate with any other SIEM solution you may use in your environment. The integration between MDIoT and Microsoft Sentinel helps a **security operations center (SOC)** team to quickly detect multi-stage attacks that often cross IT and OT boundaries.

Figure 10.6 shows native integration with Microsoft Sentinel:

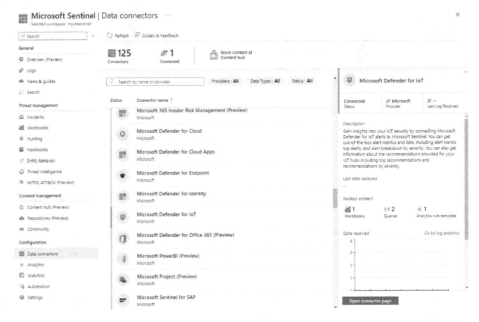

Figure 10.6 – Integration with Microsoft Sentinel using a built-in connector

The connector enables you to stream MDIoT data into Microsoft Sentinel and enables your SOC to view, analyze, and respond to MDIoT alerts and the incidents that are generated within a broader organizational threat context:

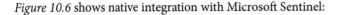

Figure 10.7 – Managing alerts from Microsoft Sentinel console

The preceding screenshot shows how you can use the Microsoft Sentinel console to handle alerts generated and take appropriate action, such as investigate, suppress, alert, or launch remediation.

You must keep three things in mind: visibility, protection, and monitoring. They help you to achieve a zero-trust objective in your OT environment.

IT organizations that develop the capability to monitor their IT/OT infrastructure and security controls in real time can enjoy great advantages. Some of the benefits of continuous monitoring are as follows:

- **Enhanced visibility and transparency of the network activity**: Real-time monitoring provides tremendous visibility into the inner workings of the IT/OT infrastructure, which is extremely beneficial to SecOps teams.

- **Rapid incident response**: Continuous monitoring eliminates the delay between the occurrence of an incident and the notification of the incident response team, enabling a quicker response to security threats or operational issues. Due to real-time security intelligence access, incident response teams can take immediate action to lessen damage and restore systems when a breach occurs.

- **Reduce outages**: MDIoT operational alerts help the operations team to react faster to operational incidents, thus avoiding unplanned downtime and rectifying errors before they lead to operations outages that negatively impact business.

In this section, we learned about how continuous monitoring makes an organization cyber-resilient, drives business performance, reduces or eliminates downtime, and helps make informed decisions and act on them in the least time.

Validating ZTA with attack vectors

ZTA implementations can be validated by creating attack vector reports. These reports give you a visual simulation of the attack vector and path to the specific asset. The following screenshot shows how a PLC may be attacked by navigating connected devices in the path:

Figure 10.8 – An attack vector report

The preceding figure shows the attack vector simulation on an important device, **PLC1**.

The path from the internet to the PLC can be achieved in three easy steps:

1. **Workstation 1** has an internet connection.

2. There are known vulnerabilities on the device that the attacker may leverage to compromise the device.

3. There is network connectivity between two subnets, meaning the attacker may enter another network. Also, there are known vulnerabilities on the PLC device, which may be used to compromise the PLC.

Simulation is an integral part of any cybersecurity program. However, given the sensitivity of the OT system (which may not tolerate any active probing), MDIoT helps an organization to do this attack vector simulation without actually attacking the device; it also provides constantly updated information on all possible attack vectors for a device in question, thus immensely helping the team to augment security.

Summary

In conclusion, we learned about the ZTA narrative with respect to the IoT/OT industry. We also looked into connecting the OT or IoT environment to further monitoring through a Microsoft SIEM solution – that is, Microsoft Sentinel. While we learned about the application and features of MDIoT with regard to OT and IoT in earlier chapters, we touched upon enterprise IoT in this one. A corporate environment consists of many IoT devices requiring visibility and monitoring, which we can cover with enterprise IoT.

We then moved on to learn more about MDIoT and how it helps in addressing the challenges of cybersecurity in Industry 4.0. We delved deeper into understanding the core features of MDIoT, such as asset inventory, continuous monitoring and vulnerability management, and threat monitoring. Finally, we completed by aligning IoT/OT security with ZTA and NIST and connecting MDIoT with a SIEM solution for better insights, visibility, and risk mitigation. Throughout this book, we hope you have enjoyed learning about cybersecurity in Industry 4.0, the Purdue model, and some of the common attacks in the IoT/OT industry.

Index

Packt.com

Subscribe to our online digital library for full access to over 7,000 books and videos, as well as industry leading tools to help you plan your personal development and advance your career. For more information, please visit our website.

Why subscribe?

- Spend less time learning and more time coding with practical eBooks and Videos from over 4,000 industry professionals

- Improve your learning with Skill Plans built especially for you

- Get a free eBook or video every month

- Fully searchable for easy access to vital information

- Copy and paste, print, and bookmark content

Did you know that Packt offers eBook versions of every book published, with PDF and ePub files available? You can upgrade to the eBook version at packt.com and as a print book customer, you are entitled to a discount on the eBook copy. Get in touch with us at customercare@packtpub.com for more details.

At www.packt.com, you can also read a collection of free technical articles, sign up for a range of free newsletters, and receive exclusive discounts and offers on Packt books and eBooks.

Other Books You May Enjoy

If you enjoyed this book, you may be interested in these other books by Packt:

Microsoft Defender for Endpoint in Depth

Paul Huijbregts, Joe Anich, Justen Graves

ISBN: 978-1-80461-546-1

- Understand the backstory of Microsoft Defender for Endpoint
- Discover different features, their applicability, and caveats
- Prepare and plan a rollout within an organization
- Explore tools and methods to successfully operationalize the product
- Implement continuous operations and improvement to your security posture
- Get to grips with the day-to-day of SecOps teams operating the product
- Deal with common issues using various techniques and tools
- Uncover commonly used commands, tips, and tricks

Practical Internet of Things Security - Second Edition

Brian Russell, Drew Van Duren

ISBN: 978-1-78862-582-1

- Discuss the need for separate security requirements based on the IoT device you're using

- Apply security engineering principles on IoT devices

- Master the operational aspects of planning, deploying, managing, monitoring, and detecting the remediation and disposal of IoT systems

- Use Blockchain solutions for IoT authenticity and integrity

- Explore additional privacy features emerging in the IoT industry, such as anonymity, tracking issues, and countermeasures.

- Design a fog computing architecture to support IoT edge analytics

- Detect and respond to IoT security incidents and compromises

Packt is searching for authors like you

If you're interested in becoming an author for Packt, please visit `authors.packtpub.com` and apply today. We have worked with thousands of developers and tech professionals, just like you, to help them share their insight with the global tech community. You can make a general application, apply for a specific hot topic that we are recruiting an author for, or submit your own idea.

Share Your Thoughts

Now you've finished *IoT and OT Security Handbook*, we'd love to hear your thoughts! Scan the QR code below to go straight to the Amazon review page for this book and share your feedback or leave a review on the site that you purchased it from.

https://packt.link/r/1804619809

Your review is important to us and the tech community and will help us make sure we're delivering excellent quality content.

Download a free PDF copy of this book

Thanks for purchasing this book!

Do you like to read on the go but are unable to carry your print books everywhere?

Is your eBook purchase not compatible with the device of your choice?

Don't worry, now with every Packt book you get a DRM-free PDF version of that book at no cost.

Read anywhere, any place, on any device. Search, copy, and paste code from your favorite technical books directly into your application.

The perks don't stop there, you can get exclusive access to discounts, newsletters, and great free content in your inbox daily

Follow these simple steps to get the benefits:

1. Scan the QR code or visit the link below

https://packt.link/free-ebook/9781804619803

2. Submit your proof of purchase

3. That's it! We'll send your free PDF and other benefits to your email directly

www.ingramcontent.com/pod-product-compliance
Lightning Source LLC
Chambersburg PA
CBHW060137060326
40690CB00018B/3915